Farm Fresh

TENNESSEE

Farm Fresh

TENNESSEE

THE GO-TO GUIDE TO GREAT Farmers' Markets
Farm Stands • Farms • U-Picks • Kids' Activities • Lodging
Dining • Wineries • Breweries • Distilleries • Festivals
and More

Paul and Angela Knipple

The University of North Carolina Press   Chapel Hill

A **SOUTHERN GATEWAYS** GUIDE

© 2013 PAUL AND ANGELA KNIPPLE. All rights reserved. Manufactured in the United States of America. Designed by Courtney Leigh Baker. Set in Whitman with Bellow and Gotham display by Rebecca Evans. The paper in this book meets the guidelines for permanence and durability of the Committee on Production Guidelines for Book Longevity of the Council on Library Resources. The University of North Carolina Press has been a member of the Green Press Initiative since 2003.

Library of Congress Cataloging-in-Publication
Knipple, Paul.
Farm Fresh Tennessee : the go-to guide to great farmers' markets, farm stands, farms, u-picks, kids' activities, lodging, dining, wineries, breweries, distilleries, festivals, and more / Paul and Angela Knipple.
pages cm. — (A Southern gateways guide)
Includes index.
ISBN 978-1-4696-0774-0 (pbk.)
1. Pick-your-own farms—Tennessee—Guidebooks. 2. Farmers' markets—Tennessee—Guidebooks. 3. Agriculture—Tennessee—Guidebooks. 4. Tennessee—Guidebooks.
I. Knipple, Paul. II. Title.
SB319.863.T2K55 2013
381'.4109768—dc23   2012041234

Southern Gateways Guide™ is a registered trademark of the University of North Carolina Press.

paper 17 16 15 14 13   5 4 3 2 1

IN MEMORY OF OUR GRANDPARENTS,
who gave us our first taste of farm life.

AND IN MEMORY OF OUR FRIEND MICHAEL LENAGAR,
who rekindled our love of the land.

CONTENTS

WEST
TENNESSEE

Lake
Obion
Weakley
Henry
Dyer
Gibson
Carroll
Benton
Crockett
Lauderdale
Haywood
Madison
Henderson
Decatur
Tipton
Chester
Shelby
Fayette
Hardeman
McNairy
Hardin

MIDDLE
TENNESSEE

Stewart
Montgomery
Robertson
Sumner
Macon
Clay
Pickett
Houston
Cheatham
Trousdale
Jackson
Overton
Fentress
Dickson
Davidson
Wilson
Smith
Putnam
Humphreys
DeKalb
White
Hickman
Williamson
Rutherford
Cannon
Perry
Maury
Warren
Van Buren
Lewis
Bedford
Coffee
Grundy
Wayne
Marshall
Moore
Sequatchie
Lawrence
Giles
Lincoln
Franklin

Scott
Campbell
Claiborne
Hancock
Sullivan
Johnson
Hawkins
Union
Grainger
Washington
Carter
Morgan
Anderson
Hamblen
Greene
Cumberland
Knox
Jefferson
Unicoi
Roane
Cocke
Loudon
Sevier
Bledsoe
Meigs
Blount
Rhea
Monroe
McMinn
Marion
Hamilton
Bradley
Polk

EAST
TENNESSEE

INTRODUCTION

We both grew up spending time in rural areas—Angela in southwest Tennessee and Paul in northwest Mississippi. We were both very familiar with our own set of small back roads, some paved, many not. Now, after having spent the last year traveling our beloved home state from end to end, our set of back roads is much larger. And we are much the better for it.

Paul grew up in Memphis, but his grandparents still lived on their family land in the country. Weekends and summers ensured that Paul got his farm time in. Though he missed the days of the farm being worked by his grandparents themselves, there were chickens to tend to, a large garden to maintain, and all the dirt a boy could ever hope to play in.

When Angela was five, her parents and grandparents decided to leave Memphis and move to the country, so her experiences were even richer. A road blocked by cows was a valid excuse for being late to school. Summer days were spent picking horned tomato worms from the vines and shelling purple hull peas on the front porch at her grandmother's knee. Tending to chickens and learning how to deal with the native wildlife were all a part of it, along with swinging on grapevines and popping tar bubbles on the road with her toes.

Things changed when college started. Angela moved back to the city, and by then, Paul's grandparents had passed away. We started to become part of what we think of as a lost generation. During our teens, farming and farm life began to undergo a radical change driven primarily by economics. We remember well Willie Nelson, John Mellencamp, and Neil Young and their creation Farm Aid. In 1985, American farmers were in danger of losing their farms due to mortgage debts.

Although Congress provided relief to farmers, change still came as farms consolidated into large-scale operations and farming became increasingly industrialized. It is this movement to greater mechanization that has led to the most significant change in farm life, the flight of people from the country to the city.

This migration to cities is by no means a new trend. Beginning with the end of the Civil War, southerners both black and white left for jobs in the industrialized North. The only significant difference today is that the southern economy is strong enough to provide jobs here. Nonetheless, people are still leaving farms for those jobs.

It is in these people and their children that we see what we think of as a lost generation. Just as modern conveniences have bred a generation of people who do not cook at home, the move away from the land is creating a disconnection between Americans and their food.

But we're not here to preach that the end is near—quite the opposite, in fact. We're not the only ones to have noticed that we are drifting away from our roots. Today, people all across Tennessee are acting to reverse this trend.

In our travels, we met young people who have chosen agriculture as a career. They are working to meet a growing demand for fresh, locally grown food. And more, they are addressing not just an external market need but an internal call to reconnect with the land.

We also met people who never lost that connection. Across the state, there are many families that have been farming the same land for 100 years—even 200 years, in some cases. Not only are these families still farming, but they are opening their farms to others to show the realities of farming and to get others interested.

It's only natural that agriculture should have a draw for Tennesseans; it's practically in our genetic makeup. For over 200 years, the state seal of Tennessee has borne two words, "Agriculture" and "Commerce." Of course, when Tennessee entered the United States, agriculture was critical for the survival of the families of the state, supplying food, clothing, and shelter.

While those needs have not gone away, they are met for most of us at the grocery store and at the mall. Nonetheless, there is a spark in the heart of Tennessee that is fueled by the land. A year spent traveling our state introduced us to a wide range of people with that spark. We hope this guide will encourage you to explore Tennessee and kindle your own spark.

HOW TO USE THIS BOOK

This book is arranged around the three Grand Divisions of Tennessee—West, Middle, and East. Each region is divided into categories of interest. If a farm could fit into more than one category, we chose the category that we thought best fit its primary business. The categories are sorted alphabetically by county, with businesses in the same county also sorted alphabetically.

Farms

All farms listed in this category are working farms, although some may be open to the public only seasonally. We've included farms that are organic as well as those practicing conventional agriculture. They are, however, all family farms. Some farms do charge a fee for tours, so you should ask in advance when scheduling your visit. Most farms also sell their goods. We recommend that you buy something when you visit a farm, if possible. Not only are you helping the farmers economically, but you also get to take home some delicious products.

Farm Stands and U-Picks

The farms listed here interact with the public through stands on or near the farms where they sell their products or by allowing visitors to come into the fields to pick their own produce. The most popular U-pick operations involve strawberries, blueberries, peaches, and apples, although you will come across some that offer other things as well. This is by no means a complete listing of all of the U-pick opportunities or farm stands available in our state.

Farmers' Markets

Farmers' markets are popping up all the time, so we know that we haven't listed all of them here. The ones we have listed, though, have mostly producer-only content, meaning that the items sold in them by the vendors are only what they grow or make. Most of these markets also have locality requirements, meaning that the vendors who sell there must reside in certain counties or live within a specified distance from the market.

Choose-and-Cut Christmas Trees

While Tennessee doesn't have a huge Christmas tree industry, the people who choose to be a part of it all seem to be in it for one reason—seeing tradition carried on through generations of families. The farms listed here allow you to go into their fields of trees and choose the one you want; most will even let you cut the tree yourself.

Wineries, Breweries, and Distilleries

During the late 1800s, vineyards were often planted on land in Tennessee that didn't look suitable for any other agricultural use, and grapes seemed set to become an important cash crop for the state. But in 1919, Prohibition all but wiped the vineyards out. Wineries are a relatively new business in Tennessee, having appeared here only within the last quarter of the twentieth century. Tennessee wine is a growing industry, and you're sure to discover wineries that we didn't list.

You'll find that we talk about the experience of visiting the wineries instead of about the wines they offer. We want you to experience those too and decide for yourself what you think of them. All of the wineries we visited offer tastings. We've also included a couple of breweries whose location is a tremendous part of why the brewery even exists. And, we can't forget that Tennessee is famous for whiskey and white lightning, so we've got some distilleries whose dedication to local ingredients can't be overlooked.

Stores

While we've listed a couple of chains of grocery stores that are doing a good job of working with local farmers, for the most part these stores are single location mom-and-pop shops. You'll find them in cities and also in the middle of nowhere.

Dining

Finding restaurants that are working with local farmers wasn't a hard task. More and more restaurateurs are seeing the benefits of using local food. What we've tried to do here is to list the best that we've tried, the ones that seem most devoted, the ones that have local food written into their core philosophy. And we've tried to give you a range of restaurants. It's easier for a fine-dining restaurant to pay a higher ingredient price than it is for a burger joint, but we've found some great burger joints that are willing to do it and still stay in the price range you'd expect for a Tennessee burger joint.

Lodging

Staying on a farm is an experience that shouldn't be missed. The fact is you can make of the experience whatever you want it to be. The lodgings we've listed here, for the most part, will let you sit back, prop your feet up, and just relax, if that's what you're looking for, or the proprietors will let you get your hands dirty and work right along with them, if that's more what you have in mind.

Special Events and Attractions

Tennessee has a wealth of agricultural fairs and festivals along with county fairs that you don't want to miss out on. We've also listed some of the great museums in the state that celebrate agriculture and some of the historical recreations of rural farm life that your entire family will enjoy.

Recipes

Each section ends with a selection of recipes generously provided by some of the people we met while writing, all featuring the freshest seasonal produce.

Appendixes, Index

At the end of the book, you'll find a glossary of agriculture terms, a listing of sites by county, information on some of the great farm-related organizations that are out there, and an index that includes every listing and recipe in the book.

Dining Price Key

$	inexpensive; most entrees under $15
$$	moderate; most entrees $15–$25
$$$	expensive; most entrees over $25
$$$$	deluxe; set price over $100

Lodging Price Key

$	inexpensive; rooms under $100
$$	moderate; rooms $100–$150
$$$	expensive; rooms $150–$200
$$$$	deluxe; rooms over $200

Let's Not Stop Here

Visiting these places and meeting the people who run them was some of the most fun we have ever had. We are thrilled at the thought of people reading this book and having adventures like we did. We want to hear about your experiences, and we want you to be able to share with others. Go to Facebook and like our page, "Paul and Angela Knipple," or visit the official website for the book, www.farmfreshtennessee.com, and tell us and others about what you found in our beautiful state.

As the Greek philosopher Heraclitus said, "Nothing endures but change." Some of these places may be gone by the time you read this. New places will have opened. We will do our best to use Facebook and the website to keep you updated about changes. We ask you to do the same. If you find an exciting new place, let us know.

THINGS TO KEEP IN MIND WHEN VISITING A FARM

To plan a visit, always contact the farm at a reasonable hour. Family farms are exactly what they sound like, a blend of home and business. You may have trouble reaching anyone during the day, but you should be able to leave a message. If you call at night, call well before bedtime, because farming is usually an early-to-bed-early-to-rise business. Even if a farm lists public hours, call to verify them before a visit.

Wear sturdy, comfortable shoes and clothes that you don't mind getting dirty. A good farm is all about developing good soil, but at the end of the day, when it is on your clothes, it is still dirt. And it's not always perfectly level dirt. Plots for growing crops can be soft or damp. Pastures where animals roam can be broken or rutted. The best protection against a twisted ankle or a fall is a good pair of shoes.

Remember that germs abound in nature. Animals leave presents, and the rich soil of a farm also teems with life. Make sure that you have hand sanitizer in the car, and do your best to keep little fingers out of little mouths.

Pack all of the things that you would for any other outside day: water, sunscreen, hats, and insect repellent. It's also a good idea to take along some reusable shopping bags and a cooler so that you'll be prepared to take home some of the great purchases you can make at the farm.

Leave your animals at home. Yes, your puppy would enjoy a chance to run and play in the fields, but most farms have dogs that may not take well to newcomers. Unfamiliar dogs can also frighten the other animals living on a farm.

Pay attention to where you and your children are going, and always listen to what the farmer tells you. Farm equipment is incredibly dangerous, and accidents can happen in an instant. You also have to be aware of where you are wandering, lest you meet up with an angry bull or a hive of highly protective honeybees.

Assume that all fences are electric fences and that all electric fences are on or "hot" until you are explicitly told otherwise.

Don't set up unrealistic expectations. Farms aren't petting zoos, and young children shouldn't go with the idea of petting cuddly animals. Not only are you setting up for an unhappy situation, but it could be a dangerous one. Fully grown farm animals can and do bite. Big pigs and bulls are really big. Never take an animal's gentleness for granted, especially where a child is involved.

West Tennessee

West Tennessee is a relatively low and flat part of the state between the Tennessee River to the east and the Mississippi River to the west. The rich alluvial soils of the region have been a boon for agriculture but not necessarily for small farms. Because the land is so flat, it is easily made into large tracts for heavily mechanized, large-scale farming. Opportunities still abound, however. Small family farms have survived, and new ones have been created.

FARMS

Tennessee Safari Park at Hillcrest Farm

"Pass through the gate and turn right at the kangaroos." Not many Tennessee farmers can say that, but Claude Conley can. While Hillcrest Farm is a working farm where cotton is grown and cattle and sheep are raised, the side of the farm most visible to visitors is Claude's passion—the Safari Park.

Claude always knew that he wanted to have exotic animals on the farm, and his first buffalo came in the late 1950s. Monkeys, zebras, and camels followed. Today, the park is a drive-through zoo where visitors can get up-close and personal with emus, ostriches, llamas, antelopes, pigs, buffalo, and more from the comfort (and safety) of their cars. The zoo also features a walk-through section where guests can meet monkeys, hyenas, peacocks, and pygmy goats.

Friendly neighbors drop by at the Tennessee Safari Park.

The main house was built in 1830 and features two large pecan trees that were planted by Claude's great-grandfather from nuts that he brought back from a trip to Texas. Claude's grandfather built the barn. "I bet he never dreamed there'd be monkeys or large spur tortoises here."

637 Conley Road, Alamo (Crockett County), 901-734-6004, www.tennesseesafaripark.com. f

Diller Dairy Farm

Diller is the sole supplier of milk to Rock Springs Dairy, a small local dairy whose milk can be purchased in stores throughout the region. At the farm, we learn that part of what makes that milk so good is the blending of Jersey and Holstein to get just the right flavor and fat content. But what makes the farm special is the Diller family. Charles Diller explains why this is a family farm with no outside workers: "I wanted my sons to grow up on a farm the way I did. It doesn't make as much money as some things, but there are more important things than money. It's good to work with animals and learn that responsibility."

108 Peavine Road, Bradford (Gibson County), 731-676-3438.

Todd Family Fun Farm

James Todd and his family began their agritourism venture with a corn maze, but they have continually added more to make their farm welcoming year-round. In spring, the farm is open for educational visits where you can learn about a farmer's life during the planting season. Bring a GPS or borrow one from the Todds and take a hike around the farm, searching out their geocaches.

In the summer, you can stop by to pick up delicious varieties of sweet corn. Kids can play on slides and monkey bars and in converted grain bins, where they can really "get into" corn and soybeans. And in the fall, the corn maze grows larger and features a new theme every year. For littler legs or anyone who might not be up to a ten-acre corn maze, the Todds provide hay bale mazes that are fun for any age.

But the farm is about so much more than just entertainment. There's a lot of history here. The farm has been in James Todd's family for three generations. In fact, part of the farm is land that James's grandfather grew up on as a sharecropper. "Kids today don't even know how to think about it. They always ask what kind of Facebook we used to have out here. I just laugh, because my father was thirteen before they even had electricity in the house. Kids just won't know about that sort of thing if we don't teach them," he tells us.

101 Tom Austin Road, Dyer (Gibson County), 731-643-6720,
www.toddfamilyfunfarm.com.

Downing Hollow Farm

At Downing Hollow, a musician and her PhD husband, Lori and Alex Greene, have settled into a happy Ma and Pa Kettle existence. As the name implies, the farm is in a lovely hollow in the hills of West Tennessee.

A creek divides the property. On one side, Lori and Alex have their fields and are constantly working to improve the land. On the other side is a building that houses a commercial kitchen and a teaching kitchen. Outside is a wood-burning oven built by one of the farm's interns.

In addition to training new farmers through internships, the farm offers another important service, acting as the center of an informal co-op by taking goods and produce from its Mennonite neighbors to area farmers' markets. But everything is not seriousness and hard work. Lori and Alex and their interns are jovial, and the farm boasts the friendliest collection of dogs anywhere.

5290 Olive Hill Road, Olive Hill (Hardin County), 731-925-6083, www.downinghollowfarm.com. ⧆ 🐦 @downinghollow

Oleo Acres

The first thing Tim Ammons is likely to show you on a visit to his farm is his smokehouse. It's a small log structure that he saved and put back together "like a jigsaw puzzle with no picture to look at." That's just one of the many ways he's preserving the traditions of the past.

Tim raises multiple heirloom varieties of sorghum, each with distinct flavor and character. He processes his sorghum the old-fashioned way, using mule power and a press made in 1864 to crush the canes and extract the juice. Other traditions involved in the process include using Belgian mares Sissy and Sassy to plow his fields and cooking the cane juice over an open fire to thicken it into the sorghum syrup we all know and love.

Preserving old traditions is not strictly for Tim's benefit. He loves nothing more than having visitors with whom he can share his knowledge.

269 McDonald Road, Stanton (Haywood County), 731-443-0059, www.oleoacresfarm.webs.com. ⧆

Pumpkins, Pines, and Ponds

There's something for everyone at Pumpkins, Pines, and Ponds. As the name suggests, you'll find pumpkins in the fall and Christmas trees in December, and there are stocked ponds for fishing, but there's more here than that. Fall is the big season on this recently opened farm that boasts a sorghum maze, bonfires, storytelling, and hayrides that let visitors catch sight of not only the tame animals that live on this beautiful farm but also the elk, deer, and occasional fox or coyote that call it home.

1800 Hurricane Hill Road, Ripley (Lauderdale County), 731-635-5050, www.tnfarmfun.com. ⧆

Agriculture in Higher Education and On-Farm Internships

Agriculture is a wide field that can cover many academic disciplines. Students with degrees in agriculture are prepared for careers in farming, animal husbandry, and landscaping. But they're also ready to develop alternative fuels, teach, write, design and improve products, protect the environment, work in food-related industries, solve problems in laboratory settings, work in national parks, and become doctors, dentists, pharmacists, and veterinarians.

Students come out of the university environment with degrees and ideas, but where do they take them next? Some go straight into the workforce, some take their knowledge back to family farms, and others participate in on-farm internships to put their ideas into practice, which can benefit both farmers and students. These internships are typically paid as room and board in exchange for labor, and that labor can be intensive. On a small, diversified farm, an intern will gain hands-on experience in every aspect of farming, from starting seeds, to plowing fields, to dealing with weeds, insects, and weather, all while getting to implement some of his or her ideas. The experience interns gain is invaluable. They make contacts in a community that can help them start a farm of their own, they see sides of farming that they may not have encountered before, and they are more prepared to decide if this is really the way they want to spend the rest of their lives.

While internships are typically taken by current college students or recent graduates, they're also increasingly taken by older adults considering a career change to farming. If you're thinking about that change, an internship is just as valuable for you as it is for a new graduate.

Tims Family Farm

Since 1969, the Tims family has been growing tomatoes in Ripley. Today, Robert and Karen Tims, the newest generation of the family on the farm, grow more varieties of tomatoes than ever along with a great range of other heirloom variety produce. And even though you'll find the family selling their bounty at farmers' markets all over the Memphis area, you shouldn't miss out on a visit to the farm to see how they grow such diversified crops. While the heart of their success is their love for the land and a desire to feed good food to the people around them, their sustainable agricultural practices and concern for quality are important as well.

110 Tims Lane, Ripley (Lauderdale County), 731-635-2636, www.timsfamilyfarm.com. ▪

Donnell Century Farm

The farm-to-fork beef operation at Donnell Century Farm is well known to chefs and home cooks all over West Tennessee. The farm also raises cotton, soybeans, corn, and hay, but Rose Ann Donnell wants the farm to raise and be known for more. That's what led her to take a portion of this seven-generation farm and create Donnell Century Farm Adventure.

One of the first signs you'll see as you pull up to the barn at Donnell Century Farm Adventure tells you, "All grocery store food was grown on a farm." It's a simple concept, but it's an easy one to take for granted unless you're familiar with a place like Donnell. The animal barnyard is designed for visitors of every age to interact with common farm animals and learn facts about each of them. There are buildings set up to demonstrate different farm jobs for older children along with plenty of activities to help younger children burn off any excess energy. A corn maze adds to the fun in the fall, while an Easter egg hunt makes the farm a destination for spring. And it all works toward Rose Ann's goal of connecting people to their agricultural past, a time when everyone grew what they ate.

3720 U.S. Highway 70 E, Jackson (Madison County), 731-424-4526, www.donnellcenturyfarm.com. ▪

Woolfolk Farms

Woolfolk Farms takes an organic approach toward educating children about farm life. "We used to have a classroom. We'd sit them down and try to teach them something. But we found out that's not what the teachers want," Scott Woolfolk says. "So now we just let them run wild and have fun. And they still learn. Most of these kids have never seen farm animals or corn growing." While Scott spends most of the year raising cattle, corn, cotton, and hay, he's opened the farm to schools and the public every fall since 2000 by creating corn mazes, offering hayrides, and planting a pumpkin patch.

526 Hartmus Lane, Jackson (Madison County), 731-423-2583, www.marvelousmaze.com. [f]

Rose Creek Village Farms

Ray Tyler, farm manager at Rose Creek Village Farms, sums up the community's spirit simply. "If we take care of the land, the land will take care of us." You'll see that philosophy put into action in every part of the farm.

According to Ray, what it's really about is the land. The topsoil from the farm was sold years before the Rose Creek community came here. To help rebuild that lost soil, Ray composts grass clippings and leaves from the city of Selmer and uses a compost tea as a fertilizer and pest repellant. In the fall, pigs work leaves into fallow fields to prepare them for the next growing season, and chicken and sheep revitalize pastures in rotation year-round. There's always a purpose for each field, even if it's not in crop production for the market.

We visited the farm on a hot summer day, so we got to see the pigs in their summer home under the trees that cover part of the farm's acreage. Even though the pigs were enjoying rooting and foraging for acorns, they were also working to clear brush from the forest floor, moving to a different section every few weeks. Chickens in portable coops provided pest control and aeration to revitalize other pastures.

This is a farm with purpose. As Ray puts it, "It's all about connecting people with their local food system." The farm grows year-round using hoop houses in winter and shade cloths in summer. Community members volunteer their time in the fields, and Ray is planning a series of on-farm banquets so that guests see, feel, and taste what this farm is all about.

You're welcome to take a tour of the farm to learn more about a truly sustainable system. You're also welcome to come and work for a day or longer in return for produce. It's a great place for children to learn, with farm tasks available for visitors of any age.

999 Lola Whitten Road, Selmer (McNairy County), 731-645-2834, www.rosecreekvillagefarms.com.

Snake Creek Boer Goats

Richard Surratt likes to call his small farm a hobby, but the more time you spend with him, the more you see how his passion for these animals makes this something much greater than a pastime. While most Boer goats are raised for meat, Richard raises his for show, and it's easy to see why his animals consistently win prizes at the national level. These are large goats; a male show goat can weigh up to 300 pounds.

High tech and low tech stand side-by-side at this farm, where goats are artificially impregnated for the best outcome but pest control is handled by hens and cats. As interesting as all of it is, the highlight for us was the farm's newest arrival: we were lucky enough to meet a kid born only a couple of hours before our visit.

3681 Old Stage Road, Adamsville (McNairy County), 731-632-3153, www.snakecreekboergoats.com.

Urban Farms

If you ever read *The Secret Garden* and wished you had a special, beautiful place like that, Urban Farms is a perfect place for you to visit. This nine-acre farm is hidden at the end of a cove in a residential neighborhood in the heart of Memphis, one of the last places you'd expect to find open land of that size, much less a farm. Behind decades of untamed kudzu and wisteria, the farm is invisible from the busy streets surrounding it. Until the landowner gave Urban Farms permission to change it, this was unused land, but now, where there was only a field of weeds, visitors can explore and learn about rows of vegetables, tanks of tilapia, beehives, worm beds, chickens, and goats.

In addition, as part of its mission of neighborhood improvement, Urban Farms operates a market in a converted gas station where residents can purchase fresh vegetables and eggs from the farm, other goods from re-gional farms, and essentials for any pantry. The market also encourages residents to get into the kitchen by offering a growing selection of cook-books as learning materials.

198 Wills Street, Memphis (Shelby County), 901-257-9627, www.bdcmemphis.org/urban_farms.html. ⨍ ⨎ @UrbanFarmsMEM

A Garden of Hope

All urban farms take shape to support a community, but sometimes that community and that farm are something very special. In downtown Memphis, you might notice a garden of colorful raised beds filled with healthy plants, and no matter how brutal the sun the garden is often filled with volunteer workers. They're there because of the community this garden supports—the staff and patients of St. Jude Children's Research Hospital. The garden is the brainchild of Chef Miles McMath, who wanted to have a source he could access daily for fresh produce and herbs. With a vacant lot across the street, a raised bed garden seemed like a simple solution. Because the garden is on the hospital campus, the time between harvest and serving becomes minimal, meaning that the produce retains the most nutrients possible. Even though it's not a garden that is open to visitors, it's one that you can feel good about just by seeing its bright colors and by knowing the great benefit that it brings to people who need it most.

Holt Family Farms

State Representative Andy Holt grew up in the city of Knoxville, but something was missing for him. Then he joined 4-H, where he raised sheep and cows, and that was the start of his life as a farmer. After spending three years on a ranch in Wyoming and completing a degree in agriculture, he now raises hogs and beef cattle on a farm near his wife's hometown.

In the fall, he opens the farm to visitors by offering a pumpkin patch and corn maze, but the primary thing he provides is an education for the children who visit. "We'll take the kids into the pumpkin patch and pull up a vine. We'll show them the roots and the plant and explain that the flowers are where the fruit comes from. And they get so excited," Holt says.

He's excited for the future as well. As his legislative duties allow, he is planning to add activities that will keep visitors at the farm year-round. A live nativity hayride will begin after Thanksgiving, U-pick strawberries will lure guests in early spring, and an apple and peach orchard will draw them during the summer.

461 Jewell Store Road, Dresden (Weakley County), 731-364-3459, www.holtfamilyfarms.com. ⓕ

FARM STANDS AND U-PICKS

Robison's Peach Orchard
Waymon Robison was a riverboat captain with a special dedication to his family. Rather than work the month-on-month-off schedule of a captain, he took one trip at a time, making sure he could be home. After his children were born, he stopped working summers to spend time with them. To make up for lost wages, he planted peach trees, trees that became a family endeavor for twenty-six years as his children worked at home tending the orchard rather than taking jobs elsewhere. Sadly, Waymon passed away recently, but his wife, Kay, still sells peaches. And now their grandchildren are taking their turn caring for the trees.

7955 Highway 100E, Jacks Creek (Chester County), 731-989-5315. f

Ed Davis Fish Farm
At most farm stands, all that's left to do is to cook your purchases and eat, but at Ed Davis Fish Farm, an extra step is required. While the Davises do raise tilapia, catfish, crawfish, and more, their main product is bait fish. While we were there, customers ranged from a father and his young son buying two dozen minnows for an afternoon at the lake to truckers picking up hundreds of pounds to deliver to bait stores. The farm stays busy, but if you call ahead and have some flexibility, the Davis family will be more than happy to show you around.

258 Salem Road, Milan (Gibson County), 731-662-7676.

Green Acres Berry Farm
Green Acres is all about strawberries. Here, you can buy strawberries fresh from the field by the quart or by the flat. You'll also taste Green Acres strawberries in many West Tennessee strawberry wines. Weekends are busy during the strawberry season, but the farm stand's covered parking provides a great place to wait in the shade until it's your turn at the counter.

158 Medina Highway, Milan (Gibson County), 731-686-1403. f

Blueberry Ridge Goat Farm
While you're picking blueberries at Blueberry Ridge, Mike and Evelyn Dixon will make you feel like family. The farm started in 2001 as a retirement project for Mike, but it quickly grew into something that the whole

Federal and State Fish Hatcheries

Go to any nursery or garden center in the spring, and you will find seedlings ready for your garden. In our travels, we visited similar operations where the seedlings are called fry and where they are destined to be food, just not in a garden. Across Tennessee, there are ten state and two federal fish hatcheries. Some hatcheries are warm/cool-water facilities raising crappie, catfish, bass, and other species destined for lakes and reservoirs. Others are cold-water, raising trout to be released in streams and rivers.

The hatcheries perform a number of important functions. The most obvious is supporting tourism by ensuring a positive fishing experience. Another is maintaining an ecological balance in the state's waterways. While hatcheries aren't technically agriculture, we still encourage you to visit them. There is a lot to be learned about the life cycles of the fish and the makeup of our lakes and rivers.

community can enjoy. Make time to visit while you're there and sample Evelyn's blueberry treat of the day. She'll even share her recipes if you ask.

915 Cox Road, Mercer (Hardeman County), 731-658-4352, www.goatsnstuff.9f.com. **f**

Coleman Farms

Peaches are a favorite summer treat in West Tennessee, but there really aren't many orchards growing them in a large quantity. At Coleman Farms, the peaches started as a hobby that has now grown into an operation with over 700 trees producing peaches throughout the month of July. The farm stand is located at the charming farmhouse next to the orchard, but to be sure to get peaches, you need to call ahead; most Coleman peaches are reserved by locals in the know.

85 Orchard Lane, Savannah (Hardin County), 731-925-9710. **f**

Culbertson Farm

Culbertson Farm has been providing pick-your-own blueberries since 1999. While blueberries are really only beginning to become a common treat in the South, Randy Culbertson knew they would take off. Randy's wife, Jean, grew up in southeast Michigan, where her father raises twenty-two acres of blueberries. Randy recalls his first visit to the farm fondly: "We were up there in August, and I started walking through the bushes. I started grazing, and I couldn't stop. And that's when I knew we had to grow some at our house. If I liked them that much, everybody else would too." While early customers weren't sure what to do with them, business at the farm has increased every year. It's no wonder. Not only are Randy's berries irresistible, the bushes are located high on a hill overlooking a creek where locals and berry pickers flock to cool down on hot summer days.

200 Gillis Road, Savannah (Hardin County), 731-925-4872.

Schrock Family Farm and Bakery

Donald Schrock met us outside his family's on-farm bakery, his apron and beard both flowing white in the warm summer breeze. Late in the week, he and his Mennonite family begin baking pies, rolls, bread, and more to sell at the Nashville Farmers' Market on Saturday. While they don't have a store set up at the farm, they can probably dig up a tasty treat or two for you. And if he has time, Donald gives an excellent tour of the farm.

281 Kee Road, Wildersville (Henderson County), 731-967-5011.

Peach World

While peaches are what you'd expect to get from Peach World—and you will get some wonderful ones here—there's a lot more to the farm than peaches. You'll find a great variety of seasonal produce, including famous Ripley tomatoes in the summer and beautiful pumpkins and gourds in the fall. In addition to their Ripley farm stand, owners Clare and Wayne Oswald maintain a booth at the Agricenter International Farmer's Market Monday through Saturday to keep their produce readily available, spring through fall. A trip to the farm is worth the drive. It's serene and peaceful out there, and Wayne or Clare will be more than happy to take you on a tour of the fields, weather permitting.

2151 Forked Deer Road, Ripley (Lauderdale County), 731-635-3035.

Reg Carmack Farm

The Carmack farm will keep you coming back from spring through fall. Strawberries are the first crop every year at the farm, and they're known throughout the area as some of the sweetest. They're followed in summer by sweet corn and then in fall by pumpkins. Best of all, perhaps, all of the produce is prepicked or U-pick.

152 Carmack Road, Ripley (Lauderdale County), 731-635-2088.

Jones Orchard

There is no shortage of options at Jones Orchard. Plantings of strawberries, plums, blackberries, peaches, nectarines, pears, and apples ensure that, as the season progresses, there is always something to pick. Lunch at the farm stand changes daily, but white beans and corn bread are a regular favorite. Lunch satisfies, whether you are fueling up for a trip into the orchards or have worked up an appetite picking. Our favorite option is to have lunch, then just buy our produce directly from the farm stand.

H. L. "Peach Orchard" Jones founded Jones Orchard in 1940, beginning the business by selling peaches door-to-door. His son Lee is the peach expert today, and Lee's son Henry runs the day-to-day operations. Lee's wife, Juanita, manages the farm stand and, more important, the kitchen. Henry's three sons and a daughter ensure that the family name will carry on at the orchard.

7170 Highway 51, Millington (Shelby County), 901-873-3150, www.jonesorchard.com. Farm stand open year-round. Dining $ ￼

Windermere Farms and Apiaries

Ken and Freida Lansing began their certified organic farm in 2000 on their North Memphis property in the valley below Lake Windermere. This rural retreat at the edge of the city has become a place where you can harvest your own produce throughout the growing season, beginning with strawberries in late spring. And it's more than just a place to go pick fruits and vegetables—the Lansings are more than happy to show visitors every aspect of the farm and love to share their knowledge of farming.

3060 Woodhills Drive, Memphis (Shelby County), 901-386-2035, www.winfarms.com. ￼

Organic, Naturally Grown, and Certification

Food labeling is supposed to make things simpler, but when it comes to how your meat and vegetables are raised, things can get confusing. The Organic Foods Production Act of 1990 required the USDA to create the National List of Allowed and Prohibited Substances, which outlines what can and cannot go into organics. While that act worked well by defining organic food, trouble arose when the government moved to define organic producers.

In 2002, the National Organic Program (NOP) was enacted, meaning that products could be called organic only if they had been raised by certified growers. The legislation favors growers of single crops and large growers with the ability to manage the paperwork and navigate the bureaucracy. Small farmers with diversified crops are at a disadvantage because each crop must be certified. Unfortunately, the regulations have also been watered down through lobbying.

The Certified Naturally Grown program was created as an alternative to the USDA program. Farmers follow the same requirements set forth in the NOP, but the record-keeping requirements are greatly simplified. A nonprofit organization defines the requirements and oversees certification.

FARMERS' MARKETS

Main Street Farmers Market

As part of Dyersburg's Main Street Project, this farmers' market has found a permanent pavilion home at the city's downtown river park. Local vendors offer seasonal produce as well as fresh honey, jams and jellies, breads, and more. While it's great to spend a summer morning at the market, be sure to check for activities at the river park to make your day complete.

335 Clark Avenue South, Dyersburg (Dyer County), 731-285-3433, www.dyerchamber.com/community/main-street-farmers-market. ◼

West Tennessee Farmers' Market

There was a great crowd when we visited the West Tennessee Farmers' Market in downtown Jackson. With one permanent pavilion devoted to Tennessee-only products, the market encourages vendors to sell only what they grow themselves. You'll find produce, meat, and breads in great variety, from the expected to the more exotic. You'll also meet people who are interested in sharing about their farms. It's easy to spend a day here, and the scheduled learning and children's activities can mean fun for everyone.

91 New Market Street, Jackson (Madison County), 731-425-8308, www.cityofjackson.net/farmersmarket. **f**

Agricenter International Farmer's Market

The Agricenter is the world's largest urban farm and research center. The farm consists of 1,000 acres of land dedicated to testing the latest in agricultural technologies. An expo center on the property allows for conferences and demonstrations. The general public can take advantage of many features of the facility. From mid-spring through fall, the "big red barn" hosts a farmers' market. In spring, local farm Jones Orchard runs the U-pick strawberry patch. And in the fall, an eight-acre corn maze makes for fun for kids of all ages.

7777 Walnut Grove Road, Memphis (Shelby County), 901-757-7777, www.agricenter.org/farmersmarket.html. **f**

Cooper-Young Community Farmers Market

The Cooper-Young Community Farmers Market in midtown Memphis is young and growing. With its central location, the market draws customers from all over the city. But its primary focus is on serving its own community. This market accepts the USDA's Supplemental Nutrition Assistance Program, enabling customers from every income bracket to take home fresh, healthy food once a week during the market season. This is also a farmer-based market; founder Lori Greene farms at Downing Hollow Farm in Olive Hill, Tennessee. In addition, the market hosts local artists and musicians as well as artisans and farmers.

1000 South Cooper Street, Memphis (Shelby County), 901-725-2221, www.cycfarmersmarket.org. **f**

Memphis Botanic Garden Farmers Market

Audubon Park and the Memphis Botanic Garden are tranquil spots amid the bustle of east Memphis. The busy farmers' market at the garden is in an especially idyllic location. A grove of tall pine trees provides shade during the hot Memphis summers, and the gently winding walk lined with vendors feels more like the forest than a part of the city. Vendors here offer a full range of produce, meats, baked goods, crafts, and more. In the garden itself, the My Big Backyard project has created a place where children can learn more about gardening and nature while playing in a space designed just for them.

750 Cherry Road, Memphis (Shelby County), 901-636-4100, www.memphisbotanicgarden.com/farmersmarket. ◼ ✚ @memphisbotanic

Memphis Farmers Market

One of the most popular places in Memphis is there only on Saturday mornings spring through fall. The Memphis Farmers Market operates from 7:00 A.M. to 1:00 P.M. every Saturday during the season in the pavilion behind the Memphis Central Station.

Over seventy vendors offer produce, meats, baked goods, cheese, arts and crafts, and more in this market where all are required to grow and produce their products within the tri-state region. Because of Memphis's location in the state, roughly half of the vendors at the Memphis Farmers Market are from Mississippi, Arkansas, or Missouri, but the energy at the market is pure Tennessee.

While you're shopping, you can enjoy live music as your children participate in the market's weekly educational activities. Local chefs provide demonstrations using the seasonal ingredients that shoppers can take home. If you have any questions about anything while you're there, look for the market's large group of volunteers, who will be more than happy to help you.

Central Station, South Front Street at G. E. Patterson Boulevard, Memphis (Shelby County), www.memphisfarmersmarket.org. ◼ ✚ @mfmfarmersmkt

Pop-Up Farmers' Markets

The farmers' markets we've chosen to feature in this book are those with regular dates and times so that you can easily plan to visit them. But they're not the only type of farmers' market out there. Keep your eyes open as you're traveling around during the summer and early fall. You'll be surprised at how often you can see pickup trucks with beds full of produce stopped beside the highway, sometimes even in cities. Markets pop up in shopping center parking lots, around the edges of town squares, and alongside the highway whenever farmers have produce that they need to sell.

These markets don't just happen during the normal growing season, either. There are great farmers' markets that open just during the winter, and you'll find them in rather unexpected places. For example, in Memphis, at least six farmers show up to sell their products on Saturday mornings in the parking lot of Tsunami Restaurant in midtown throughout the winter.

South Memphis Farmers Market

The South Memphis Revitalization Action Plan was founded jointly by the University of Memphis and St. Andrew AME Church. The first project put into place was the South Memphis Farmers Market in an effort to address a food desert in the area. Now the market is more than a source of fresh food; it's become a place where community members gather to socialize. Area children have gotten in on the act, painting a farm-inspired mural on the building that anchors the market. The farmers and other vendors who come to the market are now as much a part of the community as the people who live there. And the market has also dedicated space to a different nonprofit health provider to offer needed services to residents each week.

Mississippi Boulevard at South Parkway East, Memphis (Shelby County), www.somefm.org. ⬛ 🐦 @SoMeFM

Court Square Farmers Market

Covington has one of the prettiest town squares in West Tennessee, and just off that square, a small pavilion is the center of a bustling and friendly market. This spring and summer market opens early, and you'll want to get there early to have the best selection of the great produce these farmers bring to sell. Later in the day, there may be fewer vendors, but the pavilion is a wonderful shady spot to relax out of the sun.

111 North Main Street, Covington (Tipton County), 901-476-9727.

Martin Farmers Market

The Martin Farmers Market shows that great things can come in small packages. This modest market sets up beside the railroad tracks in downtown Martin near the University of Tennessee campus. While the farmers here offer a great variety of produce, it's the excitement of the community that makes this market a special place. We arrived in time to see the crowd gathering, waiting for the market to open, and watched how more people kept showing up as the afternoon progressed.

101 University Street, Martin (Weakley County), 731-479-2124.

CHOOSE-AND-CUT CHRISTMAS TREES

Ward Grove Tree Farm

Roy and Sue Ward first planted Christmas trees in 1986, and now their customers from the early days return with their grandchildren to carry on the tradition. More than for just buying a tree, a visit to the farm is an adventure. A hayride takes customers to the field to choose their trees. On weekends, a mule team pulls the hayride, thereby removing any reason for you to be stubborn about going to get a tree.

20 Ward Road, Beech Bluff (Madison County), 731-427-2682, www.wardgrove.com. ∎

Duncan Christmas Tree Farm

Neighbors are responsible for Duncan Christmas Tree Farm. Yes, they buy their trees there every year, but more than that, they inspired and shaped the business. The idea began when Charlie Duncan visited a neighbor who had a Christmas tree farm. "It was about the prettiest thing I had ever seen," he says. Not long after, another neighbor just happened to find a bundle of fifty white pine saplings labeled "Christmas Trees" along the side of the road, and these became the first trees Charlie planted to grow and sell. After that, he planted more and more trees. "I was really happy with the way they were shaping up, but I could tell they weren't going to be perfect. I didn't know anything about trimming the trees," he says. That's when another neighbor mentioned that her son owned a tree farm and was willing to teach Charlie how he did it.

With his craft perfected, Charlie now offers hayrides and sleigh rides out to the fields for you to choose your tree. Afterward, you can return to the store on the property, where you can warm up and sip cups of hot coffee, cocoa, or spiced cider and shop for any decorations you might need to make your tree complete. You can even have your tree flocked, if you like.

186 Hester Road, Selmer (McNairy County), 731-645-5769, www.duncanchristmastreefarm.com. **f**

WINERIES AND BREWERIES

Crown Winery

Crown Winery makes sustainability look easy. From using grapes grown on the winery's own land to supporting local farmers to becoming the only solar-powered winery east of the Rocky Mountains, Crown is a model operation. Crown opened for retail sales in 2009 and has become a favorite location for weddings and photos.

Take a tour of the winery and vineyards with owner Peter Howard and learn how and why an Englishman made his way to creating a sustainable winery in West Tennessee. And don't forget to take home a bottle of Crown's strawberry wine, an annually produced wine made to celebrate the West Tennessee Strawberry Festival from strawberries grown at nearby Green Acres Berry Farm in Milan.

Botbyl Pottery is co-located on the vineyard and provides the opportunity for visitors to take home a unique souvenir. Select an amphora handthrown by master potter Eric Botbyl at the pottery gallery, and Crown will fill the bottle with the wine of your choice before corking and sealing the amphora in ancient Roman style.

3638 East Mitchell Street, Humboldt (Gibson County), 731-784-8100, www.crownwinery.com. **f**

Paris Winery and Vineyards

You'd expect to find great wines in Paris, but you might not be looking for them when you visit the famous city's namesake in West Tennessee. The Ciarrocchi family produces traditional Italian-style wines made solely from the grapes grown on the Ciarrocchis' nine-acre vineyard. The winery began as a part of the family farm, but Diana Gunderson, the Ciarrocchis' daughter, explains that the favorable soil and climate of the area are comparable to that of Tuscany. These conditions soon encouraged her parents to grow the vineyard into a separate commercial operation that opened to the public in 2008.

There's more to the winery than just wine. Ruggero's Italian Bistro, located on-site, not only serves traditional Italian cuisine made from family recipes but also uses beef and pork from the family's farm. The store sells wine-making supplies, local treats, and Diana's herbed olive oils and red wine vinegar alongside the wine. And if you want to visit the winery but don't want to make the drive there, the winery boasts a public runway for small aircraft.

2982 Harvey Bowden Road, Paris (Henry County), 731-644-9500, www.pariswinery.com. Dining $ 🅕

Century Farm Winery

"Carl asked me if I would mind if he grew a few grapevines, just for a hobby. I was thinking he meant ten or fifteen vines, but then he planted five acres," Jo O'Cain laughs. That hobby turned into a passion, and Century Farm Winery was born. The vineyard is on land that has been in Jo's family since the 1830s and is a designated Tennessee Century Farm. While not all popular varieties of wine grapes grow in the humidity of West Tennessee, Century is still able to grow enough to use 85–90 percent of its own fruit. The tasting room is open daily, and Jo or Carl will be happy to give you a tour of the vineyard. Children are welcome at the winery; in fact, the summer Saturday evening music series was designed to be a family-friendly event on the grounds. And the only payment required at the tastings and music shows is to give D. D. the winery dog a good belly scratching.

1548 Lower Brownsville Road, Jackson (Madison County), 731-424-7437, www.centuryfarmwinery.com. 🅕

Century Farm Winery is just one of many proud farms across the state that have been in a single family for over 100 years.

Century Farms

Agriculture is an uncertain business. A sudden frost can destroy an entire season's peaches. A rainy spring can keep a farmer from planting on time. One year can be a boom and the next a total bust. Obviously, it takes a special kind of person to keep at it.

In Tennessee, there are many such people—entire families, in fact. In 1975, as part of the national bicentennial celebration, Tennessee started the Tennessee Century Farms Program. The program recognizes families who have farmed the same land for 100 years or more. So far, the program has recognized over 1,400 farms, and more than 140 of those farms have been in the same family for over 200 years.

Ghost River Brewing Company

"Great water makes great beer," Chuck Skypeck told us when we sat down to visit with him at Ghost River's downtown Memphis brewery. That's what made Chuck and his partner, Jerry Feinstone, decide to locate the brewery—the city's only craft brewery—in Memphis, with access to the city's famously good water. The water source, the Memphis Sands Aquifer, was first tapped in 1887 and has been providing the city with water ever since.

The Ghost River Wetlands area is one of the aquifer's recharge areas, a region that has been and still is in great need of protection. Ghost River Brewing Company sees the problem and donates a portion of the proceeds of every barrel of beer sold to the Wolf River Conservancy, an organization devoted to protecting and enhancing the Wolf River corridor, part of which is the Ghost River State Natural Area.

Ghost River is proud of being a local brewery, and its beer has been available only a very limited distance from the city, but it is easy to find at many bars and restaurants in the area. In fact, you couldn't buy Ghost River beer at the grocery store until 2011, when the company added a bottling line to the brewery. Saturday afternoon tours are available, but they fill up months in advance, so make your plans in accordance and reserve your space.

827 South Main Street, Memphis (Shelby County), 901-278-0140, www.ghostriverbrewing.com. f

Old Millington Vineyard and Winery

"You can make a small fortune with a winery. Well, you can if you started with a big fortune," laughs Perry Sorensen, owner of Old Millington. For Perry, the winery is definitely a business, but it's a business he loves through good and bad years. Although Old Millington doesn't grow enough vines to provide all of the grapes the winery needs, Perry is able to get fruit from other local vineyards. Sunday afternoons are the best time to visit the winery. On those afternoons, Old Millington hosts live music, and for a minimal fee, you can bring in a picnic, buy some wine, and enjoy a day among the vines.

6748 Old Millington Road, Millington (Shelby County), 901-873-4114, www.oldmillingtonwinery.com. f

STORES

China Grove Country Store
China Grove Country Store is a hub of good food in its rural community. Here, you can get the freshest milk and butter from nearby Rock Springs Dairy, produce from area farms, locally produced meats and preserves, and fresh-baked goods, whose aroma will draw you in like nothing else. But don't miss out on a special treat. For a simple, delicious lunch, pick up a sandwich from the deli made on homemade bread and pair it with an apple or muscadine cider slushy, the perfect way to cool down on a summer day.

445 China Grove Road, Rutherford (Gibson County), 731-665-7431. Dining $

Backermann's Bakery and Cheese
Backermann's has been selling baked goods and preserves in Whiteville for twenty-six years. Ten years ago, local dairy farmers Earl and Mary Yoder purchased the bakery and added a store. When we asked Earl what made him decide to take on the business, he smiled before he explained. "A young couple moved down from Ohio to our dairy farm, and he needed a job. I took him in, thought of him as my own son, so I bought the store for him to manage." They were partners in the business for four years until the young couple moved to a new Mennonite community in Van Leer and the Yoders took over the business full-time.

While the Yoders no longer operate their dairy farm, they are raising all-natural, grass-fed beef cattle. You can purchase that beef in the store along with fresh eggs and any baking ingredient you can imagine. Their baked goods are available in the store, and you can find them at the Easy Way chain of grocery stores in Memphis. When we asked him what he liked most about having the market, Earl said simply, "Being on a dairy farm, you get isolated. It's good to meet people now."

260 U.S. Highway 64, Whiteville (Hardeman County), 731-254-8473. Dining $

Tripp Country Hams

You'll find Tripp ham and bacon in stores throughout West Tennessee, but nothing beats a visit to the shop in Brownsville. Here, you can pick up extra bits and pieces for whatever you might need, along with whole slabs of bacon and Tripp's award-winning ham. USDA regulations prevent the facility from being available for a tour, but your nose will have an adventure that will get your mouth watering for more.

207 South Washington Street, Brownsville (Haywood County), 800-471-9814, www.countryhams.com.

Ada's Unusual Country Store

A lot of places claim to be unusual, but few live up to the name. Ada's Unusual Country Store is one of the few places that does. There are things you expect to see in a country store, and Ada's has all of them. But when you look closer, you'll see what we saw. Along with the local honey, baked goods, flour, herb teas, and baking mixes, there are toasted cornmeal, soy flour, bee pollen, corn silk, nettle leaves, and even nori. One of Ada's signs that we saw along the highway announced that the store sells hoop cheese and twenty other varieties. That sign was no exaggeration. Not only can you gather some unique finds to take home, but if you're hungry, Ada's will make you a very reasonably priced sandwich right at the deli counter.

9653 U.S. Highway 45 N, Bethel Springs (McNairy County), 731-934-9310, www.adascountrystore.com. Dining $

Canale's Grocery

Canale's calls itself the "Home of the Hams" with good reason. Smoked boneless ham is the specialty here, and you can take it home with you whole or by the pound. If you're too hungry to wait until you get home, pick up a fresh ham sandwich from the cooler; it may make it out of the parking lot. While ham is what Canale's is famous for, you can also pick up local produce in baskets by the door throughout the spring and summer.

10170 Raleigh LaGrange Road East, Eads (Shelby County), 901-853-9490. Dining $

Easy Way

The Easy Way chain of stores has been a Memphis institution for over eighty years. Starting with a single location downtown, the original founder's grandsons now operate seven stores, all painted bright orange to draw your attention. The stores carry a full range of produce, as much of it as possible from local vendors. The dairy case contains local cheese and milk, and you'll need that milk to wash down the scrumptious baked goods available from area bakers.

Locations in Memphis and Bartlett (Shelby County),
www.easywayproduce.com. ⛶ ⯑ @EasyWayProduce

Whole Foods Market

Nationally, Whole Foods has been working with local farmers to bring the freshest produce to its customers for twenty-five years. By supporting what is locally grown, the company puts a face behind the products and keeps a connection to seasonality, reduces environmental impact and transportation costs, preserves the agricultural heritage of the community, encourages farmers to diversify their crops, and gives farmers a maximum return on their investment. Look for the "Local" markers in stores to find produce, meat, dairy, and more that have come into the store from nearby producers.

Locations in Memphis, Nashville, Franklin, Chattanooga, and Knoxville,
www.wholefoodsmarket.com. ⛶ ⯑ @WholeFoods

DINING

Pickwick Catfish Farm Restaurant

It doesn't get much fresher than right out back. While the fish farmed behind the restaurant are harvested only once a year, they are flash-frozen to ensure freshness and quality. One thing the restaurant does differently than most is to serve primarily catfish steaks, a piece made by cutting the fish from side to side to make bone-in, inch-thick cross sections. Few fish processors sell that cut these days, since fillets can be sold at a higher price. At Pickwick, though, the owners do the cutting themselves, so they can keep serving steaks.

The most unusual—and most delicious—thing that Pickwick serves is smoked catfish. Skinned and beheaded catfish are brined overnight, rubbed down with pepper, and slow-smoked over hickory wood chips. If you're not in the area, Pickwick ships the smoked catfish, but you owe it to yourself to go and try the entire menu.

4155 Highway 57, Counce (Hardin County), 731-689-3805, www.pickwickcatfishfarm.com. $ **f**

Scott-Parker Barbeque

When you're traveling across Tennessee, you'll find barbecue everywhere you look. But, unfortunately, most of this pork comes from pigs raised and processed somewhere else. Ricky Parker of Scott-Parker Barbeque does things the old-fashioned way. First, its one of the few restaurants in Tennessee where whole hogs are smoked over open pits every day. Second, those pigs come to the restaurant from local farmers via a local slaughterhouse.

With whole hogs, you can choose which part (or parts) of the hog you want, and you can get the meat on a sandwich or by the pound. Just plan to stop by early in the day. The restaurant takes its time to produce a delicious product, but it comes in finite amounts, and when it's gone, it's gone.

10880 U.S. Highway 412, Lexington (Henderson County), 731-968-0420. $

Brooks Shaw's Old Country Store

When you think about restaurants with a desire to promote and support local producers, you probably aren't thinking about buffets. But the Old Country Store is a wonderful exception to the rule. The restaurant is proud to purchase as much as possible from local farmers. It also encourages employees to farm by offering a ready customer for their produce. You'll find traditional southern favorites on the buffet, including multiple varieties of greens, candied yams, fried chicken, and the restaurant's famous cracklin' corn bread. In the attached country store, the Shaws sell products from local producers, including Tripp Country Hams and Smith Sorghum Farm.

The restaurant is also a strong proponent for other independent restaurants in West Tennessee. In 2010, the Shaws produced the West Tennessee Culinary Map, a free guide for locals and tourists alike to find and enjoy the flavors and traditions of mom-and-pop restaurants in small towns throughout the region.

56 Casey Jones Lane, Jackson (Madison County), 731-668-1223, www.caseyjones.com. $ **f** **y** @oldcountrystore

Acre

Chef Wally Joe came a long way to become a Memphian. His family immigrated to Cleveland, Mississippi, when he was young. Once there, they owned a grocery store until his father decided to buy a local restaurant. Having grown up influenced by both the South and the cooking of his Chinese grandmother, Wally fell in love with food. He made the Cleveland restaurant a huge success before coming to Memphis.

Now Wally and his longtime kitchen cohort, Andrew Adams, have opened a spectacular restaurant in East Memphis. The menu exhibits a creative use of local products that balances their love of their native South with Wally's Asian background and other global influences.

690 South Perkins Road, Memphis (Shelby County), 901-818-2273, www.acrememphis.com. Lunch $$, dinner $$$ 🟦

Amerigo

The Amerigo chain of Italian restaurants stretches from Memphis to Nashville, and each location does its part to support local agriculture. Owner Ben Brock is a member of the board of the Memphis Farmers Market. The Middle Tennessee restaurants frequently sponsor events and provide demonstrations at the Franklin Farmers Market. Diners at the restaurants can enjoy the local meat and produce that is used as often as possible.

Locations in Memphis, Jackson, Brentwood, and Nashville, www.amerigo.net. $$ 🟦

Andrew Michael Italian Kitchen

Andrew Michael Italian Kitchen is truly a labor of love. As children, chefs Andrew Ticer and Michael Hudman fell in love with cooking while watching their Italian grandmothers create the meals they loved. Today, they share some of their grandmothers' recipes along with innovative and original takes on Italian cuisine in their East Memphis restaurant. Their dedication to providing the best quality and best flavor for their guests drives their commitment to sourcing as much as they can locally. These chefs frequent farmers' markets and purchase their meat as whole or half animals, cutting it themselves to waste as little as possible.

While they're best known in the city for their delicious pork dishes, Andy and Michael are equally adept at preparing unique vegetarian options. In their kitchen, tradition marries technology to create a dining experience like no other in the region. On the last Monday of the month,

Project Green Fork

Memphis-based Project Green Fork is devoted to sustainability in the restaurant industry. Studies show that approximately one and a half pounds of trash are produced for each restaurant meal that is served, and restaurants can compost or recycle close to 95 percent of the 50,000 pounds of garbage that each establishment produces annually. Project Green Fork certifies restaurants that want to change those numbers. Memphis-area restaurants can become certified by using sustainable products, setting up recycling programs, composting their kitchen waste, using nontoxic cleaners, completing an energy audit, and taking steps to reduce water and energy consumption and prevent pollution. All certified restaurants are proponents of local farms and are locally owned. Look for the Project Green Fork logo at restaurants to know that you're eating somewhere that cares.

No Menu Monday, the chefs offer the best of their innovations, and diners get to enjoy some of their latest and tastiest experiments at a set-menu dinner. While cuisine like theirs may seem challenging for younger diners, both chefs enjoy seeing children experience something new and will work hard to create a memorable experience for any family event.

712 West Brookhaven Circle, Memphis (Shelby County), 901-347-3569, www.andrewmichaelitaliankitchen.com. $$$ 🅕 🅨 @amitaliancooks

Brushmark

Located in the Memphis Brooks Museum of Art, Brushmark overlooks the sweeping lawns of Overton Park. Chefs Wally Joe and Andrew Adams (also of Acre) use their flair and local ingredients to bring luxurious dishes to the elegant space. The museum is also a proud supporter of local agriculture, frequently playing host to films and festivals celebrating farmers.

1934 Poplar Avenue, Memphis (Shelby County), 901-544-6200, www.brooksmuseum.org/brushmarkrestaurant/. $$ 🅕 🅨 @BrooksMuseum

Café Eclectic

Where do you go in Memphis when you want unique food made with local ingredients without spending a lot of money? One delicious answer is Café Eclectic. Since opening in 2008, Café Eclectic has been devoted to locally sourcing as much of the restaurant's ingredient list as possible. And the eatery has succeeded well. At every meal, diners can choose great options featuring local flavors. And, while they're not based on strictly local ingredients, Café Eclectic's coffee and dessert menus are filled with original creations that can satisfy any sweet tooth.

603 North McLean Boulevard, 901-725-1718; 111 Harbor Town Square, 901-590-4645; Memphis (Shelby County), www.cafeeclectic.net. $ 🛐 🐦 @cafeeclectic

The Elegant Farmer

Memphis restaurateur Mac Edwards has a longtime commitment to local food, serving as a founding board member of the Memphis Farmers Market. While Edwards is at the market working, he can also shop for his latest venture, The Elegant Farmer. In a Tudor-style building shared with an antiques shop, the restaurant offers a cozy dining room and a sunny patio where diners can enjoy refined southern classics. From its opening, the restaurant has been a success. "It's nice when doing good is good for business, but it really all comes down to doing the right thing," says Edwards.

262 South Highland Street, Memphis (Shelby County), 901-324-2221, www.theelegantfarmerrestaurant.com. Lunch $, dinner $$ 🛐 🐦 @elegantfarmertn

Felicia Suzanne's

Chef owner Felicia Willett is a vivacious highlight of the Memphis food scene, and it shows in her menu. She displays her flair and dedication to using local ingredients with dishes like her deviled eggs, made even better with the addition of house-smoked wild king salmon and a small topping of domestic caviar. You'll find pimento cheese on the menu as well as classics like her gumbo and turtle soup. Our favorite thing, though, is her southern charm, which comes through in her gentle drawl and in dishes like her BLFGT salad, a glorious tower of locally produced bacon, lettuce, and fried green tomatoes with rémoulade sauce.

80 Monroe Avenue, Suite L-1, Memphis (Shelby County), 901-523-0877, www.feliciasuzanne.com. $$$ 🛐 🐦 @feliciasuzannes

Interim

Chef Jackson Kramer has faced one problem since the day Interim opened. With a perfectly executed fine-dining menu served in an exquisitely decorated space, everyone clamors for his cheeseburger, which uses locally raised Angus beef. Jackson takes it in stride, though. The native Memphian pays homage to his southern roots with dishes like macaroni and cheese and a vegetable plate made using fresh, local, seasonal produce. He stays close to home even with his fanciest dishes, like butter-poached prawns from a nearby farm.

5040 Sanderlin Avenue, Memphis (Shelby County), 901-818-0821, www.interimrestaurant.com. Lunch $$, dinner $$$ [f]

Las Tortugas Deli Mexicana

It's not hard to imagine how wonderful local produce tastes in traditional southern cuisine, but at Las Tortugas, owner Jonathan Magallanes shows just how delicious those same products can taste in traditionally prepared Mexican cuisine. Most of Memphis seems to approve of his approach; it's not uncommon to see a line stretching into the parking lot during the lunch hour.

Unlike the Tex-Mex so many of us are used to, Jonathan serves a selection of sandwiches made on homemade bread as well as tacos and a number of side items not found anywhere else in town. The most popular is *elote*, an ear of corn rolled in crumbled cheese and chili flakes.

Jonathan is also proud of his ingredients. "No big trucks pull up here. We go out every single day and shop for the best and freshest produce," he says. He has started receiving deliveries now, though, as local farmers arrive weekly to bring him much of the beef, pork, and eggs he uses.

1215 South Germantown Road, Germantown (Shelby County), 901-751-1200, www.delimexicana.com. $ [f]

The Majestic Grille

Dinner and a show are a big part of the attraction at The Majestic Grille. Deni and Patrick Reilly's restaurant was originally a silent movie theater built in 1913. Now the interior has been lovingly restored and serves as an elegant backdrop for Patrick's elevated comfort food, much of which is made with local ingredients from the nearby farmers' market. Also close by is the historic Orpheum Theatre, which allows folks a chance to see Broadway shows or classic movies after their dinner at Majestic.

145 South Main Street, Memphis (Shelby County), 901-522-8555, www.majesticgrille.com. $$ @majesticgrille

Restaurant Iris

Restaurant Iris had a lot to live up to after moving into the building occupied by La Tourelle, a Memphis landmark for thirty years. From the beginning, chef Kelly English focused the menu around locally available products. Today, diners can find locally sourced ingredients in dishes in every course of the meal.

"The approach we take to food at Iris is fancy takes on the comfort food I grew up with," Kelly tells us. "But it really centers on the local food system that's available here. One of the great things about Tennessee is that it's so wide-stretching. In Memphis, we're literally in the northernmost part of what many southerners consider to be the South, so I get to see things here that I didn't see growing up in Louisiana, but we still have those blazing summers, and some of my favorites are favorites here too." But even more than that, he says, "cooking is all about friends, and when you're dealing with the products from local farmers, even when they're not there, you're cooking with friends."

2146 Monroe Avenue, Memphis (Shelby County), 901-590-2828, www.restaurantiris.com. $$$ @restaurantiris

Trolley Stop Market

The only thing better than farm-to-table is farmer-to-table. Arkansas farmers Jill and Keith Forrester have been coming across the Mississippi River to the Memphis Farmers Market since it opened. After a few years, they decided to make a permanent place for themselves. While they still take their fresh produce and flowers to local markets, they now operate their own restaurant and market.

Named for the trolley stop right in front of the business, the market is a convenient source of locally raised meats, and the restaurant uses local meats and produce in its dishes. Especially good are the catfish sandwich and the Margherita pizza. And if you're lucky, Jill or Keith may be the ones delivering the food to your table.

704 Madison Avenue, Memphis (Shelby County), 901-526-1361, www.trolleystopmarket.com. $

Yolo Frozen Yogurt

Frozen yogurt is a tasty treat, but the Memphis-based Yolo chain makes it especially appetizing with a wide variety of locally sourced toppings. Fruit, granola, and even the cookie crumbs and brownie bites come from area producers. Plus, Yolo shows its commitment to sustainability through its participation in Project Green Fork.

Locations in Memphis, Bartlett, Collierville, and Jackson, www.yolofroyo.com. $ 🅕 🐦 @yolofroyo

Marlo's Down Under

Considering the name, you might expect Marlo's to have a touch of Australia on the menu, but that's not what the "Down Under" is about. The restaurant is in the basement (down under) of a shop facing the town square in Covington. There's a lot of history in this space. The building was originally a grocery store owned by the Naifeh family, and the basement was used as storage. Jimmy Naifeh, longtime speaker of the house in Tennessee, left very small footprints in the concrete of the floor. Today those footprints have become part of the wall.

Owner Ron Scott has taken considerable steps to preserve the history of the building and other area landmarks. The bar in the restaurant today is the original fruit counter from the grocery store, and the stained glass behind the bar was salvaged from a long-gone Memphis restaurant, The Loft. Marlo's is a family place. Ron is there most nights greeting guests, while son Nick is the chef and son Todd mans the bar. It's a great restaurant for parents and children, serving food that can please any palate.

Ron tells us that he buys from local farmers as much as possible because it does make such a difference in the food quality. By partnering with a local cattle ranch, Marlo's is able to offer one of the best burgers around, along with steaks that melt in your mouth. The restaurant's produce comes from both local farms and the local farmers' markets. "It costs more, but it's so much better, and our customers appreciate that," says Ron.

102 East Court Square, Suite D.U., Covington (Tipton County), 901-475-1124, www.marlosdu.com. $$ 🅕 🐦 @MarlosDownUnder

LODGING

Todd Farm Bed and Breakfast

Angela died at Todd Farm. She was stabbed in the back with an ice pick. But she got better. It was all part of the fun at one of Pat and Rodney Todd's mystery dinners at their lovely bed and breakfast.

Rodney farms 270 acres of pine, and guests can enjoy over twelve miles of scenic walking trails, abundant wildlife, and a growing blueberry patch. The bed and breakfast opened in 2009 because, as Pat says, "we just like people and getting to visit with different people." And they do. What they've created is a beautiful place to rest, relax, and just get away from everything for a little while. But most of all, it's a place where you almost instantly can make yourself at home.

3945 Highway 424, Cedar Grove (Carroll County), 731-987-3730, www.toddfarmtn.com. $$ 🟦

Stillwaters Farm

A getaway to the cottage at Stillwaters Farm offers an experience like no other. Rich and Valeria Pitoni enjoy being able to provide a taste of a simpler life, and they succeed beautifully. Guests at the cottage can choose to be as much or as little involved as possible with the goings-on of the farm itself. The cottage is furnished with lovely antiques and is surrounded by lush flower gardens and perfect picnic spots that might tempt guests to stay right there. But it also allows them access to 100 acres of farmland, where the farm's friendly resident animals will interact with visitors of any age and where guests are invited to join in life on the farm.

375 Oak Grove Lane, Henderson (Chester County), 731-989-4251, www.stillwaters-tn.com. $$$ 🟦

SPECIAL EVENTS AND ATTRACTIONS

Tennessee River Freshwater Pearl Museum and Farm

You might not expect to find a freshwater pearl farm in Tennessee, and there's really no reason you should, since this is North America's only freshwater pearl farming operation. While the museum and gift shop are open daily, tours should be scheduled in advance but are worth the effort and expense. You'll get to head out to the docks to meet a local diver, see a fresh catch of mussels, and learn how those mussels are an important part of the river's aquaculture. The farm manager will also demonstrate shucking a mussel to reveal the freshwater pearl inside. The gift shop gives you the opportunity to take home a pearl, Tennessee's official state gem.

255 Marina Road, Camden (Benton County), 731-584-7880, www.tennesseeriverpearls.com.

Green Frog Village and Cotton Museum of the South

Green Frog Village is a great example of early rural life in West Tennessee. Schools bring groups of children who marvel at the old schoolhouse and cabins, but the center of it all is the Cotton Museum of the South, and the center of the museum is the Eatman Gin, a working cotton gin that exemplifies the importance of a gin to even a small cotton farmer in the early twentieth century. The village is open for tours all summer and hosts a great festival in the fall. And in good years, summer visitors may be able to pick up some of the great wild blackberries and raspberries that grow on the property.

15 Green Frog Lane, Green Frog (Crockett County), 731-663-3319, www.greenfrogtn.org.

West Tennessee Agricultural Museum

While it's not uncommon to see antique farm equipment at stores and farms across the state, there's nowhere like the West Tennessee Agricultural Museum to see farm equipment and more put into the context of history. Displays throughout the museum show different aspects of farming and rural life in West Tennessee from the early pioneer days through the present. Cotton is a major focus for both the museum and the region, and you'll be able to see how the plant has been grown and used here for over 250 years. Check the museum's calendar for special demonstrations, and

Agricultural Festivals

Agriculture is big business, and as the farmers in these pages teach us, it is a lifestyle that many would not give up. Agriculture is also a reason to celebrate. As time passes, different communities across the state hold festivals as their particular specialties come into season.

Festivals are a source of pride and a reason for locals to get together. They are also a boost to the area's economy. In the short term, tourist dollars flow in during a festival. In the long term, festivals are an important part of quality of life in an area and may help attract new business.

There is no shortage of festival topics. Some cover fruits and vegetables, both cultivated and wild. Others feature animals such as mules or goats. Some really great festivals focus on the history of a particular place or on the history of agriculture in general. Regardless of the reason for the celebration, you are sure to find welcoming people, good eats, and plenty of fun. And you just might learn something.

plan to take your family to the annual Fall Folklore Jamboree, when the museum celebrates even more aspects of rural West Tennessee life.

3 Ledbetter Gate Road, Milan (Gibson County), 731-686-8067, www.milan.tennessee.edu/museum/.

West Tennessee Strawberry Festival

The West Tennessee Strawberry Festival has celebrated the regional strawberry industry since 1934. The festival has grown since then into a week of events that includes pageants, parades, a strawberry recipe contest, a carnival, art exhibitions, music, and more. Throughout the festival grounds, you'll find vendors selling strawberry treats ranging from simple strawberry shortcake to unexpected deep fried strawberries.

1200 Main Street, Humboldt (Gibson County), 731-784-1842, www.wtsf.org. Held in May.

Lauderdale County Tomato Festival

Once June arrives, you'll notice signs all over West Tennessee announcing the arrival of Ripley tomatoes. What's so special about these tomatoes? Well, it's actually the soil in Lauderdale County that makes these some of the most flavorful tomatoes around, and the county loves to celebrate them. In the city park, you'll find arts and crafts, a carnival, local music and food, and of course tomatoes for sale—both fresh and preserved. On the town square, you can enjoy an antique car show and an annual exhibition of locally produced tomato art.

123 South Jefferson Street, Ripley (Lauderdale County), 731-635-9541, www.lauderdalecountytn.org/living_tomato.html. Held in July.

The Cotton Museum of Memphis

The farms that fed the Industrial Revolution in the United Kingdom were West Tennessee farms. Factory workers ate locally grown food, of course, but the first wave of the revolution was textiles, and the spinning wheels and looms of those mills processed cotton. Even before Memphis was a city, the bluffs along the Mississippi River were the place to sell cotton. Until the Civil War, Memphis was home to the largest and most respected cotton exchange in the United States. Even after the war, cotton was the backbone of the Memphis economy.

In 1922, the Memphis Cotton Exchange built a new office and trading floor on Front Street in downtown Memphis. Today, that building houses The Cotton Museum of Memphis. The exchange trading floor has been painstakingly restored to the way it was in the 1930s. Oral histories and exhibits document every aspect of the cotton business—from the fields to the trading floor and beyond. And if you've never felt unprocessed cotton, you can remedy that here. Hands-on exhibits let you touch different varieties and grades of this southern staple crop.

But what makes this such a special place to visit isn't just the history; it's the wonderful way that history is presented and explained in this museum. "I believe in the true Latin meaning of museum: to muse, to entice," says museum director Carol Perel.

65 Union Avenue, Memphis (Shelby County), 901-531-7826, www.memphiscottonmuseum.org.

The International Goat Days Festival is a great place to make new friends.

International Goat Days Festival

No kidding around, Goat Days is a fun festival. From the world's greatest goat parade to fainting goat shows to goat chariot races, it's all about the goats here. While local concessioners offer their own specialties, none of them offer anything featuring goat, since the guests of honor at this festival might get a little offended by that. Wander through the paddocks to meet the stars, but be careful and forewarned. Many of the breeders have goats for sale, and it's very tempting to go home with a new best friend.

4880 Navy Road, Millington (Shelby County), 901-872-4559. Held in September.

RECIPES

Memphis Summertime Tomato Salad

The summer heat in West Tennessee means two things. First, the Ripley toma-
toes are about to be ready. Second, it's time to find ways to cool off. This salad
from Karen Tims of Tims Family Farm takes advantage of fresh tomatoes to
make a refreshing salad.

SERVES 8

FOR THE DRESSING:

⅔	cup red wine vinegar
⅔	cup avocado oil
2	tablespoons ground white pepper
2	tablespoons salt

FOR THE SALAD:

1	red onion, diced
1	English cucumber
2	quarts mixed color cherry tomatoes

In a large mixing bowl, combine the dressing ingredients, whisking until
thoroughly incorporated.

Add the onion to the dressing and toss to coat. Allow the onion to
marinate in the dressing at room temperature while you prepare the
cucumber and tomatoes.

Dice the cucumber and halve the tomatoes. Add them to the dressing
mixture and stir to coat and combine thoroughly.

Refrigerate the salad for at least 30 minutes or until ready to serve.

Spicy Sweet Peach Deliciousness

Chef Kelly English of Restaurant Iris is a regular customer at farmers' markets
in Memphis because he demands the freshest ingredients. While he is serious
about his produce, he also has a playful side that shows in creative dishes like
this one.

SERVES 4

1	jalapeño pepper
4	medium peaches, peeled and pitted
¼	teaspoon red pepper flakes

Pinch of garlic powder

Pinch of powdered ginger

2   tablespoons sugar

1   lime

1   tablespoon rice vinegar

10 cilantro leaves

5   mint leaves

Salt to taste

TO SERVE:
Sweetened whipped cream

TO ROAST THE JALAPEÑO PEPPER:

If you have a gas stove, use tongs to hold the pepper directly in the fire, turning to roast all sides. Otherwise, heat a heavy skillet over medium-high heat. Cook the whole pepper in the hot skillet with no oil, turning the pepper until its skin is blackened and blistered all over.

Immediately place the pepper in a paper bag and roll the top to seal it. Allow the pepper to rest in the bag for 3 to 5 minutes. Remove the pepper from the bag and rub it gently to remove the skin.

TO ASSEMBLE THE DISH:

Remove the stem, seeds, and membranes from the jalapeño and discard them. Mince the pepper and add it to a medium mixing bowl.

Dice the peaches and add them to the bowl with the jalapeño.

Add the red pepper flakes, garlic, ginger, and sugar to the peaches, stirring to combine.

Zest the lime and juice it. Add the zest and juice to the peaches along with the vinegar.

Finely mince the cilantro leaves and add them to the peaches.

Lay the mint leaves in a stack one on top of the other. Roll them as tightly as possible lengthwise. Slice the leaves across the roll as thinly as possible. Add the slices of mint leaf to the peaches and stir to combine all ingredients. Season with salt to taste.

Cover the bowl tightly with plastic wrap and refrigerate it for at least one hour before serving.

Serve with sweetened whipped cream (or, God forbid, frozen whipped topping).

Peccorino Sformato with Cauliflower Puree and Bacon Salad

Southern chefs today are taking traditional ingredients and elevating them to great heights. Andy Ticer and Michael Hudman of Andrew Michael Italian Kitchen specialize in modern takes on Italian classics, but their ingredients are as purely southern as they are. Andy shared this recipe. While the preparation and plating make for a stunning dish, in our minds the cheese and cauliflower still hearken back to a good old casserole.

SERVES 6

FOR THE SFORMATO:

6	tablespoons butter
6	tablespoons all-purpose flour
2	cups heavy cream
¾	cup half and half
¼	teaspoon salt
¼	teaspoon freshly ground black pepper
6	eggs, separated
8 ½	ounces grated Parmesan cheese

FOR THE CAULIFLOWER PUREE:

1	head cauliflower, trimmed and roughly chopped
3	cups heavy cream
	Salt and freshly ground black pepper to taste

FOR THE BACON SALAD:

1	tablespoon olive oil
1	slice bacon, diced
8	brussels sprouts, trimmed and with the leaves plucked individually
2	tablespoons sherry vinegar
3	tablespoons celery leaves
3	tablespoons parsley leaves
2	teaspoons grated Parmesan cheese
½	teaspoon lemon zest
	Salt and freshly ground black pepper to taste

TO MAKE THE SFORMATO:

Preheat the oven to 425 degrees.

Melt the butter in a large skillet over medium heat.

Gradually add the flour, stirring constantly until it is completely blended with the butter. Cook, stirring constantly, until the mixture becomes a blonde roux—slightly darker and beginning to smell nutty.

Add the heavy cream, half and half, salt, and pepper, stirring constantly until these ingredients are completely incorporated.

Whisk the egg yolks in a bowl. Temper the egg yolks by stirring a spoon of the cream mixture into the bowl. Continue to stir in the cream mixture one spoonful at a time until the temperature of the mixture in the bowl is roughly the same as the cream mixture in the skillet. Add the tempered egg yolks to the skillet, stirring until completely incorporated.

Add the Parmesan cheese, stirring until completely incorporated.

Whip the egg whites until they achieve medium peaks. Fold the egg whites into the cream mixture.

Pour the mixture into 6 buttered 6-ounce pie tins or a buttered large muffin pan.

Bake in a water bath for 30 minutes.

Rotate then bake for an additional 15 minutes until the tops are golden brown.

TO MAKE THE CAULIFLOWER PUREE:
Add the cauliflower and cream to a large saucepan over medium heat. Bring to a simmer and cook for 15 minutes, stirring occasionally.

Strain the cauliflower from the pan and add to a blender. Puree the cauliflower until smooth.

Add salt and pepper to taste.

TO MAKE THE BACON SALAD:
Heat the olive oil in a large skillet over medium heat. Add the bacon and cook, stirring frequently, until brown.

Add the brussels sprouts to the skillet and remove it from the heat.

Deglaze the pan by adding the sherry vinegar and stirring to remove any bits stuck to the skillet.

Toss the celery leaves, parsley leaves, Parmesan, and lemon zest in the bacon mixture.

TO ASSEMBLE THE DISH:
Spread a portion of the cauliflower puree on a plate. Place a sformato on the puree. Top it with the brussels sprout salad.

Cushaw Spoonbread

When we visited Holt Family Farms, Andy Holt made sure we left with a complete sampling of his pumpkin crop. We took home pie pumpkins, fairytales, butternut squash, and a southern classic, cushaw squash. We had an excellent time cooking our way through it all. Our favorite dish was this cushaw spoonbread, which made an appearance on our Thanksgiving table.

1	small to medium cushaw squash (about 4 pounds)
1	cup corn meal
1 ¾	cups whole milk, divided
4	tablespoons butter
1	cup heavy cream
1	teaspoon cinnamon
1	teaspoon freshly ground nutmeg
½	teaspoon salt
2	tablespoons light brown sugar
3	eggs, separated

Preheat the oven to 375 degrees.

Halve the cushaw lengthwise and scoop out the seeds. Place the squash in a baking dish, cut side down. Cover it with foil and bake it for 45 minutes or until the squash can be easily pierced by a fork.

While the cushaw is cooling, whisk together the cornmeal with ¾ cup of the milk in a small mixing bowl.

In a large saucepan, melt the butter over medium heat. Add the remaining milk, the heavy cream, cinnamon, nutmeg, salt, and brown sugar. Bring the mixture to a boil, stirring frequently, and then remove it from the heat.

Scoop the cushaw flesh into a food processor and pulse until the flesh is smooth.

Add the cushaw to the milk mixture in the saucepan and stir to combine. Return the pan to medium heat and bring the mixture back to a boil. Reduce the heat to low and slowly, stirring constantly, add the cornmeal mixture. Continue cooking, stirring constantly for 15 to 20 minutes or until the cushaw mixture thickens.

Preheat the oven to 400 degrees.

Remove the mixture from the heat and allow it to cool for at least 15 minutes.

Add the egg yolks to the cooled mixture one at a time, stirring to combine between each addition.

In a large mixing bowl, beat the egg whites until they form soft peaks. Gently fold the egg whites into the pudding until just combined.

Butter a 3-quart baking dish. Gently transfer the pudding to the baking dish. Bake for 45 to 60 minutes or until the top is golden and shiny and beginning to crack.

Allow the pudding to rest for 15 minutes before serving.

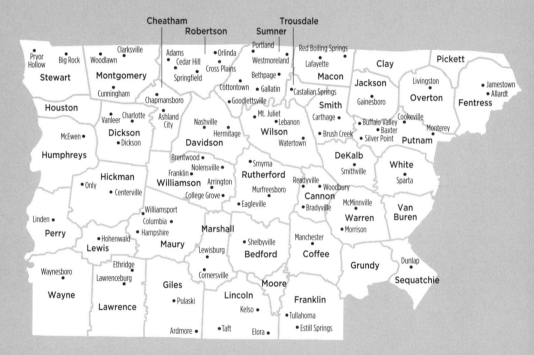

Pryor
Hollow
Big Rock
Woodlawn
Clarksville

Stewart
Montgomery

Cunningham

Houston

Vanleer
Charlotte

McEwen
Dickson

Dickson

Humphreys

Linden

Perry

Hohenwald
Lewis

Waynesboro

Wayne

Ethridge
Lawrenceburg

Lawrence

Cheatham

Robertson

Adams
Cedar Hill

Springfield

Chapmansboro

Ashland
City

Nashville

Hermitage

Davidson

Brentwood
Nolensville

Franklin

Williamson
Arrington

College Grove

Williamsport
Columbia
Hampshire

Marshall

Lewisburg

Cornersville

Giles

Pulaski

Ardmore

Only

Centerville

Hickman

Orlinda

Cross Plains

Cottontown

Goodlettsville

Mt. Juliet

Lebanon

Wilson

Watertown

Smyrna

Rutherford

Murfreesboro

Eagleville

Shelbyville

Bedford

Lincoln

Kelso

Taft

Elora

Moore

Sumner

Portland

Westmoreland

Bethpage
Gallatin

Castalian Springs

Smith

Carthage

Brush Creek

DeKalb

Smithville

Readyville
Woodbury

Cannon

Bradyville

Manchester

Coffee

Franklin

Tullahoma

Estill Springs

Trousdale

Red Boiling Springs

Lafayette

Macon

Jackson

Gainesboro

Buffalo Valley
Baxter
Silver Point

Putnam

White

Sparta

Van
Buren

Warren

Morrison

McMinnville

Grundy

Dunlap

Sequatchie

Clay

Livingston

Overton

Cookeville
Monterey

Pickett

Jamestown
Allardt

Fentress

Middle Tennessee

Middle Tennessee is a large area of rolling hills in a wide basin created by the Tennessee River to the south—in Alabama, actually—and the Cumberland River in the center of the region. The state capital, Nashville, is at the heart of food activity with exciting restaurants and thriving markets, but that's not to say that the rest of the area is quiet. Opportunities abound, from wineries in the north to fresh berries in the south.

FARMS

Owen Farm

Ty Owen will tell you that the farm he operates with his brothers is a working sheep farm, but it's so much more than that. "My grandmother owned a country store out here," Ty tells us. "So for us, customer service comes naturally." That customer service shows up in the year-round events at the farm, like the annual Easter egg hunt, concerts, and the six-week fall festival that includes a corn maze, pumpkin patch, and hayrides. Ty also loves to talk about the educational aspect of the farm. In the Jimmy Maxey Learning Center, named for a former public school agriculture teacher, displays are set up to teach children about the agricultural products of the region, including tobacco. The sheep are also a learning experience for many kids. The sheep the Owens raise are meat sheep because "sheep are smaller than cows, so the kids could help out more," Ty laughs.

825 Crocker Road, Chapmansboro (Cheatham County), 615-642-0294, www.owenfarm.com. ⬛

Bells Bend Neighborhood Farms

When we met Eric Wooldridge at Sulphur Creek Farm—one of the four farms that make up the Bells Bend project—we knew right away that Bells Bend was a special place. Eric is part of the new generation of farmers, graduates of university agriculture and sustainability programs who are devoting their lives to the land.

In Bells Bend, that devotion shows as work progresses on the four complementary farms in the community. While Eric manages the farms, their existence is community-driven, with projects like deer fences and storage buildings becoming neighborhood efforts in labor and materials.

The Bells Bend Neighborhood Farms project took off in 2008 when 175 acres were set aside for that purpose. The effort was made in the face of increasing pressure for urban development. It's not that the residents of Bells Bend are against development; they simply want to see sustainable development.

You can visit the farm for scheduled tours, but keep an eye on Facebook for the classes and workshops that the farm hosts.

5188 Old Hickory Boulevard, Nashville (Davidson County), 615-974-2388, www.bellsbendfarms.com. ⬛

Clover Bell Farm

Stuart and Linda Reeve raise their herd of Dexter cattle on twenty acres in Vanleer. The Dexter breed originated in Ireland and is known for being hardy and intelligent. But what you'll notice about the cattle first is that they're small. A full-grown cow may be no more than three feet tall at shoulder height, so be prepared to fall in love with little doe-eyed cows and calves when you go to visit. If you're thinking about a cow for a suburban farm, a Dexter might be the perfect choice for you, and the Reeves can help you choose exactly which you should take home.

1280 Bell Hollow Road, Vanleer (Dickson County), 615-763-6367, www.cloverbellfarm.com. ⬛ 🐦 @CloverBellFarm

Keller's Corny Country

Bryan and Tonya Keller ran into a problem that many farmers face: they got out of the beef business but didn't know what else to do with their farm. They knew they didn't want to sell it. The Kellers decided that since they had been hosting annual harvest festivals for the local community for years, they would open the farm as an educational agritourism business in

October. First they grew pumpkins, and then they added a corn maze and more attractions each year. But first and foremost, Keller's is still a farm, and all of the pumpkins and gourds sold here are grown here.

542 Firetower Road, Dickson (Dickson County), 615-441-4872, www.kellerscornycountry.com. **f**

Three Creeks Farm

Making a small farm profitable is a large concern for most farmers, but at Three Creeks, Steve and Beth Shafer have found ways to make that profit while still doing what they love. The pair raise goats, sheep, and chickens, although sheep are the primary focus of the farm. Beth shears them and makes yarn from their fleece that she then weaves into woolen goods. She happily offers demonstrations at the farm and also travels to fairs and festivals. Steve often travels with her, demonstrating his skills as a blacksmith.

The sheep are but the woolly tip of the iceberg. A visit to the farm can become anything you want to make of it. Talk to Beth about raising, harvesting, and using dye plants to color cloth. Or learn and practice the difference between spinning on a wheel and spinning with a drop spindle. Let Steve teach you about fainting goats or how to bathe chickens for shows. Talk to them about herd management and health.

Educating is natural for both of the Shafers, and it shows. Beth has much experience in public speaking, and Steve is a retired second-grade teacher. While they enjoy showing school groups the workings of the farm, they're truly at home doing the same for visitors of any age.

365 Peabody Road, Charlotte (Dickson County), 615-789-5943, www.3creeksfarm.com. **f**

Grandaddy's Farm

Since 2005, Grandaddy's Farm has opened to tourists in the fall. The rest of the year, the farm is planted with row crops, but when the weather cools down, an educationally themed corn maze and plenty of other activities draw guests from near and far. On the farm, Andrew Dixon and his family grow over seventy-five varieties of pumpkins, winter squash, and gourds that they sell in their farm market along with mums, honey, and more—all produced on the farm. And don't miss out on the homemade pumpkin pie that sells out fast every weekend.

Andrew grew up on the farm and tells us that he enjoys sharing it with children who might never have seen a real farm before. As a fourth-

Let the signs lead the way so you don't miss out on any of the fun at Grandaddy's Farm.

generation farmer, he's found all of the great things that kids will want to see, and he loves showing them off. When we were there on a sunny Wednesday afternoon, every child came off of the hayride with a huge pumpkin cut fresh from the pumpkin patch and an even huger smile.

454 Highland Ridge Road, Estill Springs (Franklin County), 931-327-4080, www.grandaddysfarm.com. ⓕ ⓨ @adixon250

Beaverdam Creek Farm

The Lingo family is one of many who are choosing a slower lifestyle that brings them closer to the land by farming. Originally from suburban Atlanta, the Lingos moved to rural Tennessee in 2008 and founded Beaverdam Creek Farm on seventy-three acres of rolling hills, pasture, and flowing water.

This is a small family farm with great aspirations. The Lingos don't cut corners and are intimately involved in every detail of their farm. They grow over 135 varieties of gourmet vegetables, fruits, herbs, and flowers that go into their Community Supported Agriculture (CSA) baskets. They're also raising a herd of grass-fed cattle.

The family aspect of this farm is perhaps the most important part of all. While the Lingo children have now grown into young adults, they've found their place on the farm, bringing new ideas, techniques, and their own talents to further the farm's goals.

516 Sulphur Creek Road, Centerville (Hickman County), 931-623-3732, www.beaverdamcreekfarm.com. 🄵

Holiday Acres Farm

Debbie Brown describes her Amish country farm's mission simply: "We're all about education here. We want children to learn something that they'll take away from being here." Debbie and her husband, Tom, have created a farm-to-market learning center and select a Tennessee-education-related theme for their corn maze every year. When we visited, the theme was the state bird, the mockingbird, and children had to answer trivia questions about the bird in order to proceed through the maze and reach a kid-sized nest that they could investigate at the end. There's also a petting zoo that offers kids the opportunity to pet and feed farm animals and learn about the ways they contribute to the farm.

346 Campbellsville Pike, Ethridge (Lawrence County), 931-829-2660, www.holidayacresfarm.com.

McDonald Farms

As we arrived at McDonald Farms to visit free-range chickens laying eggs, the first creature to greet us was a piglet in a harness. The farm's owner, Will McDonald, explained that the pigs he keeps near the chicken operation make up his breeding stock. While this piglet's brothers and sisters were sent to the pasture to fatten up for slaughter, this lucky little one was chosen by Will's daughter as a pet.

Will is rightly proud of his operation. His hens are free to roam over a large area. They are well tended and a bit spoiled—they all come running to meet Will at dinner time. Will also explains how eggs are cleaned and prepared as well as how his meat birds are harvested. Though it is a relatively small operation, there is no shortage of "doing things right."

430 Woodland Road, Hohenwald (Lewis County), 931-796-7921.

True to his nickname, Jeff Poppen, the "Barefoot Farmer," leaves footprints at his farm.

Long Hungry Creek Farm

It's nearly impossible to sum up a day spent with Jeff Poppen at Long Hungry Creek Farm. Jeff, also known as the "Barefoot Farmer," is a biodynamic farmer. He believes deeply in the connection of all life, explaining the way plants need the living organisms in the soil and then give nutrients back to the soil when they finish their cycle. After even a brief visit with him, you'll leave feeling a sense of wonder and connectedness with nature. We started our visit to the farm with a trip up to the hayloft, where we sat and talked about farming and about life, all the while looking out over Jeff's beautiful rows of vegetables.

Biodynamic Agriculture

Biodynamics is one of the most interesting and intense approaches to sustainable agriculture. Initially proposed in 1924 by Austrian philosopher Rudolf Steiner, the idea is that the entire farm is a single organism. A proper biodynamic farm includes both animals and plants so that it can, in essence, feed itself. Manure from the animals and plant wastes are composted to replenish the soil.

Biodynamics has been labeled a pseudoscience because of some of its more unusual tenets. Steiner suggested that farmers tie their work to the larger world, including the phases of the moon and the position of the planets. However, Steiner openly admitted that he had not tested such ideas and called for experimentation.

Studies have proven that the approach has its merits. Biodynamic farms have lower yields, but they also have higher energy efficiency and an overall positive impact on the environment. Regardless of skeptics, Steiner's suggested practices are in place on farms in dozens of countries around the world, including here in Tennessee.

On a walk through the property, we learned from Jeff about dry-land farming—the preparation and care for the soil in order to support plant life without irrigation. We talked about his use of chickens and his encouragement of beneficial insects as his sole methods of pest control. We also got to see one aspect of his farm that few, if any, other farmers have—a natural limestone cave where he keeps his produce perfectly cool in any weather until it goes to his CSA customers.

Long Hungry Road, Red Boiling Springs (Macon County), 615-699-2493, www.barefootfarmer.com. f

Hohenwald Elephant Sanctuary

There are many things you might expect to see in rural Middle Tennessee. Elephants, however, are probably not among them. And yet, the Hohenwald Elephant Sanctuary has made Tennessee a home for old, sick, or needy elephants where they can enjoy open pastures and forests. This is the nation's largest natural habitat refuge for elephants, and its 2,700 acres provide ample room for the elephants to roam.

So, why Tennessee? Most important, the temperatures in Hohenwald are consistent with those in the elephants' natural habitats. This climate means that the elephants can enjoy year-round vegetation, long growing seasons, and few days when temperatures are cold enough to force them to stay indoors.

This is a true sanctuary, and as such, it is not open to the public. There are opportunities, however, for you to be involved. The sanctuary operates a welcome center in downtown Hohenwald on Thursdays, Fridays, and the third Saturday of every month from 11:00 to 4:00. There are also scheduled volunteer days that allow volunteers over the age of fifteen to help out with the maintenance of the sanctuary. You aren't likely to see an elephant on-site, but the creatures can be viewed from the comfort of your own home: the sanctuary offers "Elecams," streaming video from across the habitat.

Andrews Spring Farm

Andrews Spring has something for everyone. In the spring and summer, trails are open for horse riding, and the animals in the petting zoo are eager to greet children at birthday parties. As fall rolls around, the ghosts and ghouls in the haunted corn maze begin to act up. Come Thanksgiving, the farm's carefully tended Christmas trees await a chance to go home with visitors.

1452 New Columbia Highway, Lewisburg (Marshall County), 931-652-2199, www.andrewsspringfarm.com. 🅵

Ring Farm

Since 2004, John and Thelma Ring have opened their farm to visitors every fall. While they own 130 acres, they farm 850, raising corn, wheat, and soybeans. Their corn maze focuses on Tennessee history, teaching children new lessons every year. When we visited, the maze, commemorating the 150th anniversary of the Civil War, was a map of the state with thirty-one battle sites for maze walkers to find. The Rings also plant an annual cotton maze, where children and adults can walk through and pick cotton, most likely for the first time. You'll also find slides, bonfires, wagon rides, pumpkins, and a corn cannon, which are sure to make spending a day at Ring Farm an adventure.

2628 Greens Mill Road, Columbia (Maury County), 931-486-2395, www.ringfarm.com. **f**

Patterson Place Farm and Zoo

When Emily Sleigh-Albright's father died in 1999, developers were ready and willing to turn the family farm into yet another subdivision. But that wasn't what Emily wanted. Instead, she and her husband decided to maintain the farm in her family's traditional fashion but with a unique twist. Emily's husband raises tobacco, as did her father, and Emily raises animals—but not just any animals. While you'll certainly see the creatures you expect to find on a farm—like James the pig; Joy, Thumper, Sarah, and Lily the goats; and a coop of chickens and turkeys—there are some surprises. Emily got Gordon the camel as a baby and bottle-fed him, and he loves to surprise visitors with kisses from his surprisingly soft lips. There's also Gus the zedonk (a rare crossbreed of a donkey and a zebra) and a friendly emu that shows off her unusual drumming call. A former schoolteacher, Emily enjoys educating visitors of all ages about her fuzzy family. "I've just always loved animals, and my parents tried to get them for me. They're my babies."

2480 Patterson Road, Woodlawn (Montgomery County), 931-553-0639.

Carr Ranch Wild Horse and Burro Center

Tennessee is famous for horses, with the Tennessee walking horse being a popular choice for riding and horse shows. Carr Ranch is making a name for Tennessee with another type of horse, the wild mustang of the western United States.

Paula and Randall Carr work with the Bureau of Land Management to prevent overpopulation of wild horses and burros that might otherwise be at risk of injury or illness. The Carrs accomplish this by bringing both wild and rescued mustangs and burros to Tennessee for adoption. At their facility, they give the animals time to acclimate to their new surroundings. They get a checkup, and their hooves are tended to. Carr Ranch should definitely be considered by anyone looking to buy a horse, and it also makes for a good visit by the casual horse lover.

4844 Couts Carr Road, Cross Plains (Robertson County), 615-654-2180, www.carrranch.com.

Gammon Family Dairy

In stores across Middle Tennessee, you'll see Gammon Family Dairy milk in jugs labeled "Tennessee Real Milk." But while the dairy has been operating for over forty years, the Gammons have been marketing their milk only since 2010. You can also pick it up in Chase's Corner Store (named for a young family member) at the farm. And, if you call in advance to schedule it, the Gammons will be more than happy to give you a tour of the dairy and farm.

5766 Highland Road, Orlinda (Robertson County), 615-654-8621, www.tripplecreek.com/tennrlmilk.

Lucky Ladd Farms

"I'm a city girl, and Jason is a country boy," Amy Ladd says. "One day I asked him if I could have a pot-bellied pig, so he got me one. After a while, he decided I needed sheep and goats, too. Now we have over a hundred animals." Those animals have become the business of the farm, with Jason and Amy focusing on breeding. The farm is open every weekend during the summer. Visitors can interact with animals ranging from chickens to llamas and learn facts about the different animals they come in contact with. In the fall, a sorghum maze provides additional fun, and the sorghum eventually becomes animal feed.

4374 Rocky Glade Road, Eagleville (Rutherford County), 615-274-3786, www.luckyladdfarms.com. [f]

Walden Farm

You can have a great fall day at Walden Farm, and if you're looking for a pumpkin, you've come to the right place. There are pumpkins, winter squash, and gourds here as far as the eye can see, but if you would prefer,

The animals wander into the people section of Poultry Hollow for a bit of rest and relaxation.

you can take a hayride out to the pumpkin patch to pick your own. There is also a host of friendly farm animals that children and adults can get to know, live music, a corn maze, and great food. While you can get your standard burger and fries here, you can also enjoy pumpkin fudge, pumpkin cookies, and toasted pumpkin seeds.

8653 Rocky Fork Road, Smyrna (Rutherford County), 615-220-2918, www.waldenfarm.biz. **f**

Poultry Hollow Hatchery

You don't always know where a country road will lead you, just like Todd Rutigliano didn't know where his first chicken would lead him. Ultimately, that first bird, a seventeen-year-old's hobby, led him to leave his corporate job and a life of heavy travel for a chance to be home doing what he loves. Along with his mother, Judy Wood, Todd now has sixty breeds and over 6,000 birds. They sell birds for both eggs and meat and sell eggs as well. There is no better way to learn about the birds than to tag along with Judy as she works, although it is tempting to join the ducks taking a dip in the family swimming pool.

116 Wilkerson Hollow Lane, Brush Creek (Smith County), 615-318-9036, www.poultryhollow.org. **f**

Hydroponics

When one talks about a return to the land, the subject of hydroponics is an interesting anomaly. Hydroponics is the science of growing plants without soil. Rather, a plant's roots are kept in water that contains all the nutrients needed for the plant.

When farmers talk about improving the soil, they are referring to increasing the nutrients contained in it and about ensuring that the soil is loose for roots to spread. Soil, however, is only a holding place for those nutrients. It is water, provided either through rain or irrigation, that dissolves those nutrients and delivers them through the roots.

Hydroponics works in one of two ways, circulating or non-circulating. In a circulating system, the water is pumped so that it flows over the plant roots. The water passes through a station where it is monitored and adjusted for proper nutrient levels and aerated so that the plants get needed oxygen. In a noncirculating system, plants are put in as simple a container as a jar or pot. The nutrient levels are adjusted manually, and the roots are left half out of the water so they can take in oxygen. All hydroponics operations of significant size use circulating systems to reduce labor demands.

Hydroponics has several advantages. Produce can be grown even in places where the soil is especially bad. Because the system is entirely self-contained, there is no harmful runoff of nutrients. Hydroponics systems are run in greenhouses, so it is easier to keep out pests and keep in beneficial bugs, and with no soil, plant diseases are far less likely.

The biggest complaint that people have about hydroponic produce is that the nutrients are still chemical, not natural, like those found in well-tended organic soil. Hydroponics has its supporters as well, due to its increased growing season and reduced environmental impact compared to traditional farming.

Long Hollow Suri Alpacas

After spending some time with Karl and Jan Heinrich at Long Hollow, you'll want to be an alpaca farmer too. The alpacas here are of the Suri breed, a breed that makes up only 3 percent of the entire alpaca population. Over nine years, the Heinrichs have grown their herd to seventy-five animals, all friendly and each with a distinct personality.

You can tour the farm to see the animals as well as the environmental protections put in place. Over two-thirds of the land here is devoted to wild life. Solar energy is used to light the barns and cool them in the summer, while a wood stove produces steam heat to keep the alpacas warm in the winter.

You can also tour the mill to see how the fleece is combed and spun to make yarn and learn what makes Suri fleece so special. After your tour, you can buy some yarn or even a finished product to take home with you.

698 Wallace Road, Gallatin (Sumner County), 615-452-7852, www.longhollowalpacas.com.

Walnut Hills Farm

You can buy beef, pork, and chicken from Walnut Hills Farm at several of the farmers' markets around Nashville. Your best bet, however, is to visit owners Doug and Sue Bagwell at the farm to see the pastures where their chicken, pigs, and cattle roam freely. The farm operates without chemicals, and the animals eat as nature intended. The Limousin cattle graze on grass and hay. The pigs root to their hearts' delight and feast on the plentiful walnuts come fall.

6635 Highway 231 N, Bethpage (Sumner County), 615-374-4575, www.walnuthillsfarm.com.

Farmer Brown's Hydroponic Gardens

With an engineering background and bit of a stubborn streak, J. B. Brown found himself with a never-ending science project: hydroponic gardening. Brown has three 30- × 100-foot greenhouses, one used for lettuces and another for tomatoes, cucumbers, and peppers. Brown bought the material for his first greenhouse and put it up himself. The intense labor and the steep learning curve required taught him the value of paying for an experienced construction crew.

After the greenhouses were in place, a nutrient mismeasure killed an entire crop, and a crop of tomatoes was too closely planted so that much of the fruit could not be harvested. Brown looks back on his experiences and laughs. "That's just part of it. You learn as you go along, and I'm sure I'll do something else one of these days."

Perhaps the best time to visit is during cucumber season. The prolific plants grow ten to fourteen inches a day and must be harvested twice a day. "Cucumbers that are too small in the morning will be ready by evening, and the ones that are too small then will be ready to go the next morning. They're work keeping up with," he tells us. When Brown began to grow more than he could sell, he and his wife, Ann, built a commercial kitchen, where she makes pickles, preserves, and cakes for local restaurants.

678 Arch Cope Road, Morrison (Warren County), 931-607-3446, www.farmerbrownsgardens.com. **f**

Bonnie Blue Farm and Log Cabin Stays

Jim and Gayle Tanner left central California to retire to 330 acres in Tennessee, but their definition of retirement is not a life of leisure. When they left California, they brought with them eleven goats and Gayle's talent for making goat cheese. Since they arrived, their herd has grown and so has their cheese-making operation. They now sell at area farmers' markets, and several restaurants feature their award-winning cheeses. They welcome visitors to the farm for a tour or to stay in the elegant cabin that Jim built for them to live in when they first arrived.

257 Dry Creek Road, Waynesboro (Wayne County), 931-722-4628, www.bonniebluefarm.com. **f**

Amazin' Acres of Fun

Jimmy and Karen McCulley primarily raise beef cattle, but in the spring they offer strawberries, and in the fall they open the farm as a corn maze. For the past twelve years, the McCulleys have used the maze as a way to teach area children something about agriculture while letting them have fun. The theme of 2011's main maze was "Know your farmer, know your food." As children and adults worked through the maze depicting the motto, they learned facts of farming life that would help them reach the end.

Outside of the maze, the McCulleys have set up a learning center about honeybees and beekeeping with an empty hive and tools on display. For

Gayle Tanner demonstrates part of the "glamorous" life of a farmstead cheese maker.

children who want to get hands-on with farm work, there's a milking "cow" demonstration that is rarely lacking waiting hands. The farm's petting zoo introduces children to goats, sheep, rabbits, chickens, and calves. And of course there's a hayride out to the farm's pumpkin patch.

2857 Old Kentucky Road North, Sparta (White County), 931-761-2971, www.amazinacres.com.

Blackbird Heritage Farm
When we pulled up at Blackbird Heritage Farm, we were met by the three unofficial greeting sheep huddling by the porch to avoid the light rain. Later, as we walked around under clearing skies, piglets slipped under the fence around their pasture and frolicked around us. "It's like the inmates are running the asylum," laughed owner Sherri Roddick. Sherri and her husband, Andy, and their business partner, Jamie, run a certified organic farm specializing in heritage breeds of animals.

Their dedication to these animals is tangible. The Red Wattle hogs, one of the breeds they raise, are so critically endangered that Sherri and Andy had a difficult time finding breeding stock. Now those hogs, along with cows, sheep, turkeys, geese, and chickens, all have a happy home. Still, the point is meat, and the main attraction is the artisan sausages and cured meats that the Roddicks sell at local farmers' markets.

6422 Arno College Grove Road, College Grove (Williamson County), 615-473-9555, www.blackbirdheritage.com.

Gentry's Farm

Allen Gentry is a sixth-generation farmer of the Gentry family to run this farm, and part of the seventh generation is farming alongside him. The Gentrys raise beef cattle with no hormones or unnecessary antibiotics in a farm-to-fork operation on 400 acres just outside of Franklin. In October, they open the farm to the public with a corn maze, pumpkin patch, and hayrides on farm wagons designed for families and young children. The farm's old tobacco barn has become an education center primarily used for a summer day camp.

Allen's concept of what farming should be carries through to every aspect of his farm. "We just do all we can do as a family. We got out of the tobacco business because of the negative health impacts, and those concerns carry over today. We don't spray pesticides on the pumpkin patch or the corn because people, especially kids, are going to be out there."

1974 New Highway 96 West, Franklin (Williamson County), 615-794-4368, www.gentryfarm.com.

Hatcher Family Dairy

When you see Hatcher Family Dairy milk in stores all over Middle Tennessee, it's hard to think of the dairy as really being a family farm. But that's a mistake. A member of the Hatcher family is involved in every step of the process, from feeding the cows to managing the dairy store on the farm and marketing the milk to stores in the region.

The Hatchers have been dairy farmers on the same land since 1831, but it wasn't until 2007 that the family began processing and pasteurizing their own milk. A tour of the farm will take you through every step, letting you meet cows and see the milking parlors, pastures, and the processing plant where the milk is pasteurized, bottled, and dated for sale.

6561 Arno Road, College Grove (Williamson County), 615-368-3405, www.hatcherfamilydairy.com.

Hatcher Family Dairy takes advantage of its processing plant to share family news with the world.

Real Food Farms

Real Food Farms was once an abandoned cattle pasture, but now it's being restored through the dedicated efforts of David Daily and John Cahill. They're using all the pieces of the farm to make improvements. There are "brush" goats to clear undergrowth and expand the plantable area. The farm's heritage breed American Guinea hogs root up the soil, eat weeds, and fertilize while growing fat and happy. Movable coops give chickens and guineas a safe place to rest and lay eggs and also allow them access to different areas of the farm, where they do their part by eating insects and aerating the soil.

The farm now has five acres in production year-round, filling CSA baskets for loyal customers and fans. But a Real Food CSA probably isn't what you're thinking it is. Instead of receiving a mystery box of vegetables every week, Real Food allows customers to fill their boxes with whatever is available. That way, if no one at your house eats radishes, radishes won't go to waste.

And then there's perhaps one of Real Food's greatest contributions to the community. Students from nearby Belmont College come to the farm for workdays, weeding the fields, feeding the animals, and planting seedlings. In return, Real Food offers special lower-cost CSAs for college students that keep their needs in mind. Whether they have access to a full kitchen or not, students can enjoy fresh produce that they had a hand in growing.

A visit to Real Food can become anything you want it to be. You're welcome to just look around or ask questions from these knowledgeable farmers about sustainability. And of course, you're welcome to get your hands dirty. There's always work to be done, and no hands are too small to get started.

6740 Manley Lane, Brentwood (Williamson County), 615-604-3886, www.realfoodfarms.com. ⬛ 🐦 @RealFoodFarms

Tap Root Farm

Susan Ingraham was finishing up a day of summer camp for kids when we arrived at Tap Root Farm. They were having snack time—enjoying a hot-from-the-oven homemade biscuit spread with honey from the farm's bees—and going over the things they'd learned that day. Each child received a biscuit half, but there was a plate of quartered biscuits waiting for the fastest answers to Susan's questions. Biscuits are amazing memory aids.

But even without the biscuits, a visit to Tap Root Farm is an amazing experience. While there's plenty of walking and talking, there's also learning and working that's gauged to be appropriate for tourists of any age. The farm also teaches classes on various topics like canning and freezing, horseback riding, gardening, and more.

4104 Clovercroft Road, Franklin (Williamson County), 615-594-3210, www.taprootfarm.com. ⬛ 🐦 @TapRootFarm

Triple L Ranch

Businessman-turned-cattle rancher Wallace Lee decided to move to the country when his children were young because he wanted them to learn a work ethic. "We only planned to raise kids here when we bought the place. We knew nothing about farming when we started," Wallace told us, laughing.

Over the years, the ranch has doubled in size to 900 acres. And now, in addition to grandkids, they raise registered Polled Hereford cattle along with a few sheep and goats at his daughter's Bedford Creek Goat Farm.

By having a registered herd, Wallace can trace the ancestry of his cows all the way back to 1760 England. He hadn't planned on having a registered herd, but he had to get a registered cow for his daughter's 4-H project, and the idea just grew from there. Now he's able to use his herd to offer the area something unique.

"Hereford beef tastes different. It has a different texture and flavor than what you buy at the grocery store. People like it. At first, when they would tell me that, I thought they were crazy, but then Hereford was the only beef I was eating. I tried some of that other beef, and people were right."

The ranch is a true family business. Four out of five of Wallace's children now live and work on the farm, raising their families there to learn the same work ethic that was taught to them.

5121 Bedford Creek Road, Franklin (Williamson County), 615-799-2823, www.lllranch.com.

FARM STANDS AND U-PICKS

Valley Home Farm

Park beside the lovingly restored 1835 farmhouse and walk around back to the Strawberry Shop at Valley Home Farm. From there, you can head into the fields to pick your own strawberries or blueberries in season. Beyond the berry fields, you'll catch peeks of the rest of this 350-acre farm that has become home not only to Nancy Potts but also to her brother and two sisters and their families.

If it's not the right time of year for picking fruit, the shop still has lots to offer. Nancy keeps forty colonies of bees, bottles her honey, and makes beeswax products. Her sisters produce jams, jellies, and strawberry cakes, but the best treats of all are the strawberry honey pops, popsicles made only from Valley Home strawberries and honey.

310 Potts Road, Wartrace (Bedford County), 931-389-6470, www.valleyhomefarm.com.

Farmers and Social Media

When we were young, we both watched *Hee Haw* with our grandparents. As budding food lovers, one of our favorite segments on the show began, "Hey Grandpa, what's for supper?" The rest of the cast, offscreen, would shout this at Grandpa Jones, and he would recite the most scrumptious list of traditional southern dishes. Farmers today are able to use social media to do something very similar. With Facebook and Twitter, they can list the items they will be bringing to the farmers' market or announce that they still have availability in their CSA. Farms can publicize special events, like the opening of a corn maze or the start of a spring festival, or announce closings due to bad weather. And sometimes—the best times, really—the communications aren't about business at all. Instead, they are about everyday life on the farm and joyous happenings like new baby animals. It is then that social media become less about business and more about family.

Four Valley Farm

Clarissa Barnes met her Norwegian husband, Gunnar, in perhaps a less-than-obvious place, Thailand, when both were on mission trips to Southeast Asia. They stayed abroad until their children reached college age and then returned to Clarissa's home state, Tennessee. Gunnar had worked in agricultural development and decided to raise cattle and then blueberries on their land. Now they have two large patches, one on a hillside overlooking their home and the other in a scenic valley a bit farther back. Now only their welcome is sweeter than their berries.

1405 Old Number One Road, Dickson (Dickson County), 615-763-6324.

J & J Organic Blueberry Farm

While many blueberry farmers have decided against using pesticides, Joyce and Jackson Graham have taken their farm a step further and made it certified organic. When you pick blueberries or blackberries here, you know that this farm has passed rigorous standards for that certification. And while the berries are the main focus of the farm, they're not the only thing you'll find there. When we stopped by, a beautiful patch of pumpkins was ripening, and other produce is available by season.

7152 Rock Creek Road, Tullahoma (Franklin County), 931-455-6855.

Country Tones Greenhouse at Limoland Farm

You'll find a little bit of everything at Country Tones. As you drive through the farm to reach the greenhouse, you'll see cattle and donkeys, and once you reach it, you'll discover that you have a choice: you can purchase prepicked strawberries in May, or you can borrow a colorful sun hat and pick your own. Either way, visiting the farm is a great experience. Country Tones can also help you take that experience home: you can purchase ready-to-plant garden and bedding plants at the greenhouse. And don't forget to come back in October, when the farm offers a pumpkin patch.

8076 Columbia Highway, Pulaski (Giles County), 931-363-5744.

Johnson Farm

If you were like us, you spent a lot of childhood summer days picking blackberries from wild bushes on vacant lots, being scratched by briars, bitten by chiggers, and eating more berries than you took home. At Johnson Farm, you can relive those memories without the briars and chiggers. The Johnsons offer U-pick or prepicked blackberries during the summer and recommend that you come to visit them early in the morning or later in the evening so that you're not out picking berries in the heat of the day.

2334 Columbia Highway, Pulaski (Giles County), 931-363-3490.

Bratton Farms

Bratton Farms presents an exciting opportunity because farms dedicated to U-pick greens are becoming more and more uncommon. The farmhouse here was built in the 1840s, and current owner David Bratton is part of the fifth generation of his family to farm this land. His wife, Bonnie, and their farm dog, Huckleberry, greet people when they come to pick greens. "I'm a people person," Bonnie says. "I always had a nagging feeling that I should be doing this. When I was laid off from my sales job, I knew it was time to be on the farm."

1914 Bratton Lane, Williamsport (Hickman County), 931-583-0033.

Hidden Springs Orchard

Brinna Spaetgens's grandfather established Hidden Springs Orchard in 1979 and instilled in her a love for growing plants. At Hidden Springs, U-pick or prepicked blueberries are available in summer, and then in fall you'll find a very unusual crop—Hardy kiwis. These thumb-sized fruits are something that we saw only at Hidden Springs, and Brinna is justifiably proud of this unique offering.

2204 Spring Creek Road, Cookeville (Jackson County), 931-529-1166, www.hiddenspringsorchard.com. f

Blueberries on the Buffalo Farm

Dan and Debbie Eiser offer what has to be one of the most idyllic blueberry picking experiences anywhere. Their well-spaced and mowed bushes invite you to spend plenty of time with the berries, but that's not all they have here. Come out and spend a day, and you can enjoy a mile and a half of mown walking trails, a fish pond with feed available, and a lovely gazebo overlooking the Buffalo River that's a perfect spot for a picnic. And if you can't spend a whole day at the farm, just talk to Dan and Debbie—they're wonderful sources of information about the area and are sure to provide you with ideas for more fun you can have.

78 Cemetery Road, Lawrenceburg (Lawrence County), 931-964-4578.

Dennison's Family Farm

At the Dennisons' farm stand, you'll be able to buy delicious strawberries in May and a huge variety of produce and flowers throughout the summer. Strawberries are the single main crop for this family farm, though, and during the season, you'll see the Dennisons' strawberry huts along highways all across Middle Tennessee—just give the farm a call for locations. The Dennisons are well-practiced in raising strawberries; they've been farming 650 acres for over thirty years. If you live in the area, you might also consider signing up for their CSA so that you won't miss out on any of their great produce from mid-June through the end of August.

98 Milner Switch Road, Elora (Lincoln County), 931-937-8162, www.dennisonsfarm.com. f

Russell Farms Organic Blueberries

Kirsten Russell's main business is raising alpacas, just across the border in Alabama, where she lives. But she also has a patch of land in Tennessee, where she's been raising two acres of organic blueberries since 2001. While Kirsten cares for the alpacas, her charming mother proudly welcomes people coming to pick blueberries in the well-tended patch. She'll teach you the best way to choose your berries and send you home with the knowledge that while organic may take more work, the fruits of their labor are worth all the effort.

1209 Ardmore Highway, Taft (Lincoln County), 256-828-9006.

Lone Hickory Blueberry Farm

Picking blueberries at Lone Hickory is like stepping into an enchanted forest where fruit is waiting to be picked at every turn. The bushes here are fully mature and well spaced to make a labyrinthine path through the cool shade. If you want to wander beyond the bushes, you're welcome to stroll down to the rushing creek below. No matter how leisurely you choose to make your stay, you're sure to leave with a bucket full of plump, sweet blueberries to enjoy.

1800 Tanyard Road, Lafayette (Macon County), 615-666-7167.

Bee Sweet Berry Farm

Bee Sweet hasn't been around all that long; Ron and Judi Grennier opened the farm for picking in 2010, but you wouldn't think that from seeing their large bushes and well-spaced patches. Along with blueberries that just seem to keep finding their way into your mouth instead of your bucket as you pick, they offer blackberries, raspberries, and cut flowers. And if you don't want to pick your own, they keep prepicked berries in their shop for you to take home. Also in the shop, you'll find Judi's homemade jams, jellies, and sachets.

1442 Globe Road, Lewisburg (Marshall County), 931-359-2157, www.beesweetberryfarm.com.

Forgie's Fruit Farm

Since 1994, this family-owned-and-operated farm has been tending an orchard of 1,100 peach and cherry trees. While the Forgies offer prepicked peaches in their storefront, they also give you the opportunity to get out in the orchard and pick your own. With eight different varieties, you'll be able to go back for more throughout the harvest season. And if you just can't decide whether you want to pick your own or not, you can sit down with a cup of ice cream to plan your attack. For the cherries, though, you'd best be quick to get there. The 2011 crop vanished from the trees in only five days.

2000 Collier Road, Lewisburg (Marshall County), 931-359-0153, www.forgiefruitfarm.com. **f**

Highland Berries and Produce

The Barn at Highland Berries and Produce is a great farm stand. Throughout the summer months, you'll be able to choose from some of the sixteen different blueberry varieties that are grown at Highland. In addition, the stand provides fresh Tennessee-grown produce from nearby farms as well as pesticide- and herbicide-free Amish-grown produce that will carry into the autumn months. There's always a great variety of produce to choose from, and it changes every week, so you'll want to stop by often to see what's new.

4411 Hampshire Pike, Hampshire (Maury County), 931-285-2543.

Cash and Carry Barn

At first glance, Cash and Carry's largest crop seems to be kitschy cement statuary, but don't let that first glance fool you. Inside, you'll find fresh produce, mostly grown on the farm at the back of the property, along with locally produced meats and canned goods. To be honest, you'll find a little bit of everything at this farm stand because, really, it's more of a wonderland than a store.

951 Lafayette Road, Clarksville (Montgomery County), 931-503-9921.

Berry Ridge Farms

David Webb didn't intend to have a U-pick berry farm. But the first blueberry bushes he planted produced more than his family could eat, and as they sold the fruit to friends, word got out and requests for more started coming in. Today, Berry Ridge offers over three acres of berry bushes for you to enjoy. While you're encouraged to pick your own berries for the experience, Berry Ridge does offer prepicked berries as well.

2380 Highway 111, Livingston (Overton County), 931-823-2829.

Apple Crest Farm

Of course you'll find apples at Apple Crest Farm, but you'll also find peaches and berries earlier in the summer. Dried fruits and preserves made from the orchard and honey from the bees that pollinate it are also for sale. This is a very well tended orchard that offers both prepicked and U-pick fruit. The farm also trusts your good nature by working on the honor system for payment.

14381 Tuckers Ridge Road, Silver Point (Putnam County), 615-735-7309. ⨍

Hurricane Hollow Apple Orchard

Hurricane Hollow is a farm stand you will love to visit. Edwina Boyd offers several delicious types of apples, including some very rare heirloom varieties, all straight from the orchard's trees. She'll quickly (and tastily) teach you that you've never really had a good apple if you've never had one straight from the tree. And the trees will make you glad this is a farm stand and not a U-pick, growing as they do up the very steep slope that makes the side of the hollow. It's far better to just relax at the stand and enjoy the beautiful view, like you will your apples.

4956 Medley Amonette Road, Buffalo Valley (Putnam County), 931-858-2445. ⨍

MMKM Produce at Cockspur Farm

For eight years, members of the Jeffers family have been selling their produce from their roadside stand that's become more than just a farm stand. Of course, they sell melons from their fields, but if you're on your way to a picnic, you probably would prefer to have an ice-cold melon to refresh your palate. They have you covered, with melons chilling in their refrigerators at the stand throughout the summer. The family also sells cold, fresh milk from a local Kentucky dairy, ice cream by the scoop, and a huge variety of preserves and are planning now to expand into an attached restaurant, bringing the family farm straight to the table.

8272 Burgess Falls Road, Baxter (Putnam County), 931-432-3276.

Adams Garden

Adam doesn't own Adams Garden; rather, the garden is in Adams, Tennessee, near the larger town of Cedar Hill. Owner Don Hall tells us, "Most folks don't know where Adams is, so the name still doesn't help them find me." Don and his small farm are worth seeking out, though. Here, you can pick a wide variety of produce throughout the growing season, beginning in early spring with asparagus and lasting through berry seasons, figs, persimmons, and fall produce.

7254 U.S. Highway 41 North, Cedar Hill (Robertson County), 615-696-2652.

Pumpkin Place

Pumpkins filling antique farm wagons are always a lovely fall sight. But it's even better when the wagons are lined up in front of the fields where the pumpkins grow. At Pumpkin Place, acres of green fields dotted with orange behemoths serve as the backdrop for this seasonal farm stand.

Pumpkin Place is not a year-round venture. The owner, Stephen Freeland, is a full-time schoolteacher who began raising pumpkins for the fall as a hobby. Twelve years later, he's raising more than ever. And these aren't your standard orange orbs. You'll find pumpkins in every shade of orange, yellow, green, and white and some that show them all. Some are small enough to fit in the palm of your hand, while others are too much for one person to lift.

6944 Highway 25 East, Cross Plains (Robertson County), 615-495-9720.

Shade Tree Farm

A visit to Shade Tree is a great educational experience for children and adults alike. With a reservation, you can tour the farm and orchard, where you'll meet the farm's animals, pick apples from the trees, learn how to make cider the old-fashioned way with a hand-cranked press, taste that freshest of ciders, and then learn about the history of apple pioneer Johnny Appleseed and how he spread apples across the United States.

Even if you don't have time to take a tour, you can learn a lot from a visit here. Owner Tom Head is knowledgeable and passionate about his field. With his great personality, you won't be able to help but let him introduce you to a new apple variety that you've never had or reintroduce you to a more common variety that you might take for granted. And of course, you can't leave without some fresh-pressed cider.

2087 Kinneys Road, Adams (Robertson County), 615-417-2915, www.head2thefarm.com. **f**

Woodall's Strawberries

Woodall's name is a little misleading. While you'll certainly find strawberries here in May, you'll find blueberries in June and watermelons in July. While all of Woodall's produce is available prepicked, the real fun is in getting out in the field and picking your own.

4452 Kinneys School Road, Cedar Hill (Robertson County), 615-513-4439.

Blankenship Farms

Four generations of Blankenships have farmed their land. Head of the family Steve Blankenship has been a part of the farm as long as he can remember. "I'm eighty years old," he told us, "and I've farmed all my life. My daddy started me out milking when I was five."

That hard-working spirit is why you'll get to visit the farm only in the fall, when the Blankenships sell their pumpkin crop. The rest of the year, the farm is devoted to row crops like corn, soybeans, and wheat with less cultivatable land used for beef cattle. But those pumpkins give the family a great way to be involved with the community, something Mr. Blankenship feels is especially important for children who aren't familiar with farming. "Most people just don't realize that everything they eat comes from dirt," he says.

The Blankenships offer a huge variety of pumpkins, some only decorative but others both beautiful and edible. Mr. Blankenship pointed out one of his favorites to us, the cushaw, a large crook-necked, green-striped pumpkin with yellow flesh that he told us tastes like sweet potatoes.

While stopping in just to pick up a pumpkin and chat is fun, don't miss out on the family's hay bale maze, pumpkin bowling, and a ten-acre corn maze.

5658 Halls Hill Pike, Murfreesboro (Rutherford County), 615-533-8566.

Red Rome Beauties hide in the dappled shade at Wheeler's Orchard.

The Blueberry Patch

The Blueberry Patch is much more than just a farm; it's an escape. Owner Angie Kleinau doesn't look to the farm for income. Instead, she told us that she keeps The Blueberry Patch going for the people who've kept coming back for the twenty-five years that the farm has been open. While you won't find Angie or her staff hovering over their guests, if you need help or advice, just ask and they'll be there with the knowledge that only years of experience can bring. And rumor has it that these are some of the largest berries you'll find anywhere, so expect your buckets to fill up quickly.

5942 West Gum Road, Murfreesboro (Rutherford County), 615-893-7940.

Wheeler's Orchard

A visit to Wheeler's Orchard is about so much more than a huge selection of apples. Of course, the apples are beautiful and delicious, but Wade Wheeler Jr. is the true attraction. He and his father planted the orchard in 1978 and used their apples mostly for their famous cider until the government requirement for cider pasteurization forced them, like so many other small orchards, to get out of the cider business. Regulars at local festivals, the Wheelers are also keepers of local history. Their home was built around a one-room log cabin, preserving it as a room in their home, saving the history of the people who were on this land before them. Take the time to visit, and you'll learn about a lot more than apples here.

956 Wheeler Road, Dunlap (Sequatchie County), 423-949-4255.

Bussell Berries

It may seem like a long drive out to the Bussell farm, but just keep going, and you'll get to a beautiful setting among rolling hills that will keep you coming back all summer long. In May, you can pick strawberries or just buy some that are ready for you to eat. Later in the summer, you can come back for sweet corn and a special crop of blackberries. What makes the blackberries special? "Those are the kids' crop," their father states proudly.

3 Rogers Lane, Carthage (Smith County), 615-735-9193.

Dillehay Farms

The Dillehays started farming their land in 1965, but it wasn't until 2006 that they began a farm stand operation to sell produce directly to the public. Since then, their on-site market has grown, allowing them to offer bedding plants in spring and fresh produce in summer through fall. A caring eye for quality has created a loyal customer base for them, and their location near Cordell Hull Lake draws campers looking for fresh melons and tomatoes for picnics.

14 Kempville Highway, Carthage (Smith County), 615-774-3688.

Dark Fire Farms

The name of this farm refers to the dark fire-cured tobacco that used to be grown and prepared here. But today, Ross and Jane Bagwell prefer to raise seasonal produce, daylilies, and sorghum. A visit to the farm may well send you home with not only a greater appreciation of the beauty of the region but also fresh fruit and vegetables, lilies for your garden, and a jar of sorghum for your table.

134 Howell Road, Big Rock (Stewart County), 931-232-5746.

Bottom View Farm

The Cook family has made Bottom View Farm a special destination almost any time of the year. Berries draw people to the farm from late spring through early summer, when peaches and then apples come into season. The farm's annual fall festival takes place every weekend in October and brings visitors to a pumpkin patch with train rides, donkey rides, and other activities for children. But even when there's nothing else going on, the farm offers Old Time Fixin's Restaurant, a breakfast-only venue, and the

Inside Scoop for sandwiches, burgers, and any ice cream treat you might desire.

185 Wilkerson Lane, Portland (Sumner County), 615-325-7017, www.bottomviewfarm.com. Dining $ [f]

Bradley Kountry Acres

Mike and Cathy Bradley owned and operated a dairy farm until 1996, when they decided to leave the dairy business behind and move into the fast-paced world of strawberries. Each spring, they open over five acres of strawberry fields for U-pick and sell bedding plants from their greenhouses. By the time summer rolls around, U-pick blackberries and peaches are ready, along with tomatoes and other seasonal produce. And fall sees the Bradleys offering pumpkins, gourds, and mums.

Try to keep the first Sunday in May open on your calendar for a visit to Bradley Kountry Acres. That's the date of the Bradleys' annual spring open house, where they officially kick off the strawberry season, give away great door prizes, and tempt your taste buds with fresh strawberry desserts made with strawberries picked straight from the fields.

650 Jake Link Road, Cottontown (Sumner County), 615-325-2836, www.bradleykountryacres.com. [f]

Crafton Farms

Located at the side of Highway 52 in Portland, Crafton Farms is a very popular place on Saturday mornings in May. You can join the crowds in the strawberry field or chat with the Craftons at their wagon that's always well stocked with prepicked berries. In July, you can come back and stock up on sweet corn.

209 Crafton Road, Portland (Sumner County), 615-969-6264.

The Garden on Long Hollow Pike

"If it isn't good, just let us know, and we'll make it right" is the slogan at The Garden on Long Hollow Pike. This farm stand offers summer produce as well as U-pick greens during cooler weather. And if you need more incentive to stop, there's a small paddock beside the stand with goats and a donkey for you and your children to enjoy.

3806 Long Hollow Pike, Goodlettsville (Sumner County), 615-504-6545.

Madison Creek Farms shows that farming and beauty go hand in hand.

Madison Creek Farms

If Tennessee had royal gardens, Madison Creek Farms would be one. The farm was the former home of Loretta Lynn and the place where she wrote "Don't Come Home A' Drinkin' (With Lovin' on Your Mind)," her first of many number one country hits. Now the farm is occupied by one of Lynn's daughters, Peggy, and her husband, Mark Marchetti. They started the farm to raise flowers, but when the economic downturn meant less money was being spent by consumers on flowers, they shifted some of their farm to produce. They now offer a CSA and produce at their farm stand. Visitors can also pick their own flowers, visit with the menagerie of farm animals, or just dip their toes in the creek.

1228 Willis Branch Road, Goodlettsville (Sumner County), 615-448-6207, www.madisoncreekfarms.com. @madisoncreekfms

Red Chief Orchard

The Bumbalough family planted Red Chief Orchard in 1987 after Les Bumbalough bought land that had been used for tobacco and dairy farming. While the orchard is named for the Red Chief variety of apple, apples are not the only fruit grown here. You'll find several varieties of peaches in early summer with tomatoes following. The apples start to arrive in July with tart cooking apples and keep going through October as other varieties come into season. You can get prepicked fruit and preserves in the barn, but the stars here are the apple and peach fried pies and apple cider. Choose a pie and a glass of cider and relax in the shade of the barn's wide porch, and you'll never want to leave.

2400 Hartsville Pike, Gallatin (Sumner County), 615-452-1516, www.redchieforchard.com. ⓕ

Rocky Mound Berry Farm

If you believe the journey is the destination, follow your GPS and your heart to Rocky Mound Berry Farm. Our GPS took us through the hollers on roads that make you question just how narrow a two-lane road can be. We crossed one-lane bridges, carefully lining our tires up with the wooden planks. If you believe the farm is the destination, follow the directions online for an easier route. Either way, the trip is worth it to get the opportunity to pick some beautiful strawberries in May or peaches and tomatoes later in the summer.

412 Hawkins Road, Westmoreland (Sumner County), 615-670-1432. ⓕ

Kelley's Berry Farm

It's easy to find Kelley's Berry Farm once you know what you're looking for. At the edge of the highway, a very contemporary building with a swirling shape serves as a unique farm office. In the fields beyond that building, Kelley's offers berries from May through July with strawberries in May, blackberries in June, and blueberries in July. While you're encouraged to pick your own, prepicked berries are available.

50 Riverview Estates Lane, Castalian Springs (Trousdale County), 615-633-7447. ⓕ

Randall Walker Farms and Viola Valley Berry Farm

While Randall Walker Farms is best known as a nursery, it also grows produce and sells only the produce it grows at its Morrison sales yard. When we stopped by in the fall, Randall Walker Farms had a good selection of pumpkins, peppers, and tomatillos along with jellies, jams, and ciders from its nearby Viola Valley Berry Farm. While the berries are in season, Viola Valley offers U-pick blueberries, blackberries, and raspberries.

8240 Manchester Highway, Morrison (Warren County), 931-635-9535, www.rwfarms.com. **f**

Boyd Mill Farm

History runs deep at Boyd Mill Farm. Carol Warren and her family live in the original mill operator's home that was built in 1821. The driveway is flanked by well-tended, thornless blackberry vines that one visitor called "city picking." After picking your berries, you can cool off in the shade and visit the remains of the original mill.

The time is ripe for fun before the berries are ready, too, with the family's Blackberry Jam, an annual music festival held on the farm. "We thought it would be fun to have our neighbors play on our front porch, then folks started showing up to listen, and it just exploded," says Carol. Now the event happens the week before the farm opens for picking, because, as Carol says, "music makes the berries grow." Aside from being a fun time for all ages, Blackberry Jam also raises funds for a different local charity every year.

3218 Boxley Valley Road, Franklin (Williamson County), 615-794-3867, www.boydmillfarm.com. **f**

Golden Bell Blueberry Farm

David and Tina DeBoer took over Golden Bell Blueberry Farm in 2003, but the blueberries here have been part of the Franklin community for over twenty years. The mature bushes are spaced out and well maintained, and you'll fill your buckets (and mouths) quickly. And don't forget to pet Lucky the blueberry dog at some point during your visit.

4080 Clovercroft Road, Franklin (Williamson County), 615-794-3758, www.goldenbellfarm.com.

Blackberries are the draw now, but the remains of the original mill at Boyd Mill Farm still stand.

Morning Glory Orchard

Curt and Tina Wideman's small, ten-acre orchard is open July through October, offering fresh peaches and apples, with U-pick apples in September and October if the crop is available. You'll also find a selection of fresh produce and farm-made jams and jellies. So how did the Widemans end up with an orchard? "We had originally planned a Christmas tree farm," Tina says, "but we decided we didn't like the cold."

7690 Nolensville Road, Nolensville (Williamson County), 615-395-4088, www.morninggloryorchard.com.

Circle S Farms

Circle S has operated as a family farm since 1838. Today, the Steed families open the farm to the public each spring with U-pick strawberries that draw repeat customers every year. While the strawberries are the main business of the farm, there's no reason to stop coming out as strawberry season draws to a close. The Steeds also offer summer produce along with a nice selection of homemade preserves at their farm stand.

1627 East Old Laguardo Road, Lebanon (Wilson County), 615-210-8145, www.tncirclesfarms.com. ▪

Lester Farms

There are few things more tempting than the sight of fresh watermelons resting in the shade of a farm stand on a hot summer day. You'll find those and more at Lester Farms. The Lesters sell the produce they raise on their ten acres, along with fresh eggs, preserves, honey, and homemade cookies, if you get there before they're gone. The Lesters also offer something unusual that you might want to take home—their very own barbecue sauce.

2822 Coles Ferry Pike, Lebanon (Wilson County), 615-564-0871, www.lesterfarmstn.com.

Pumpkin Hill

Terri and Mack Moss have operated Pumpkin Hill for over twenty years, but Mack freely admits that the whole thing was Terri's idea. They both grew up watching and helping their grandfathers work in gardens, so the idea of planting something wasn't novel. They started with half an acre of pumpkins, and although the first five years were rough, the business has been growing. They've recently added a petting zoo and a small corn maze.

The Mosses raise beef cattle the rest of the year, and Mack remembers how he got started in the business. "When we were ten or twelve years old, my brothers and I had our own bank accounts. We didn't have much money in them, but we had to have enough to buy a calf and buy it feed." For Mack, that first investment has grown into a 200-acre farm; you can take a long hayride across it to Pumpkin Hill's U-pick pumpkin patch.

861 Benders Ferry Road, Mt. Juliet (Wilson County), 615-758-5364. ▪

FARMERS' MARKETS

Cannon County Farmers Market

This open-air market sets up in the parking lot of the Arts Center of Cannon County in Woodbury. The center serves this rural area with theater, music, art galleries, crafts and folklore, and children's education programs and houses a restaurant emphasizing locally produced food that is open before all performances. The farmers' market offers seasonal produce along with other locally raised and produced products.

The best part about this market is that all the farmers look like characters. And they'll give you recipes and advice. As one of them told us about his offerings, "I try to get a good mess of greens in every one of these bags. This'll make you a good pot."

1424 John Bragg Highway, Woodbury (Cannon County), 615-563-2554, www.cannoncountyfarmersmarket.com. ⬛

Ashland City Farmers Market

The slogan at the Ashland City Farmers Market is "Buy Local, Eat Fresh," and that's exactly what you'll do when you shop there. This small market sets up in a park just outside of downtown and stays busy from the moment it opens. The fresh produce here is arranged in some of the prettiest displays we've seen anywhere, which makes simply browsing in the market fun.

Main Street at Washington Street, Ashland City (Cheatham County), 615-792-4420.

East Nashville Farmers Market

We visited the East Nashville Farmers Market on a day when the summer sun decided to break all the records. It was hot, the kind of hot when people who don't have to be outside aren't. But the heat didn't seem to be affecting the market. It was busy, with vendors and customers all taking part. The weekly yoga class met in a shaded corner, live music played on, and vendors offered everything from pottery and crafts to meat, cheese, milk, produce, and, perhaps the most popular treat of the day, Italian ice.

210 South 10th Street, Nashville (Davidson County), 615-585-1294, www.eastnashvillemarket.com. ⬛ 🐦 @enfarmersmkt

Farmers' Markets and SNAP

Farmers' markets have traditionally operated as cash-only or barter operations, and most in rural areas still do. But urban markets are making changes in payment options for shoppers from every economic spectrum. Fresh produce is necessary as part of a healthy diet, but for people on SNAP (the Supplemental Nutrition Assistance Program), formerly the food stamp program, the cash to shop with at a farmers' market wasn't available. At the opposite end of the spectrum, few farmers or markets have been able to accept credit cards as payment. Now, many urban markets are using a token system. Shoppers with SNAP cards or credit cards purchase tokens from the market's central counter that they use for purchases at the market. Many of the markets are able to double the tokens given to SNAP customers through grant programs to support healthier eating for low-income urban residents. By using the same token system for both credit card and SNAP shoppers, the old stigma of using food stamps is removed from the shoppers' interactions with the farmers, and their increased buying power means that more money goes home with the farmers and more healthy food goes onto tables where it is most needed.

Forest Hills UMC Farmers Market

The market stands under a massive old oak tree behind the church. Vendors sell produce, baked goods, meat, and ready-to-cook meals. There are even made-to-order omelets for those who haven't had breakfast yet. Local musicians provide entertainment from a nearby stand of smaller trees. Even on a rainy day, this market outside the hustle and bustle of Nashville was busy with shoppers loading up on purchases and visiting with the vendors and each other.

1250 Old Hickory Boulevard, Brentwood (Davidson County), 615-376-8013. **f**

Nashville Farmers' Market

It would be easy to spend an entire day at the Nashville Farmers' Market and would be impossible to leave without finding something to tempt you. This huge market is divided into three sections. The flea market offers bargains of all sorts. The market house features restaurants, a coffee shop, an international market, and vendors selling handcrafted goods. The "Farm Side" is home to locally owned businesses, including a growing number of farmers and producers selling their own wares.

900 Rosa Parks Boulevard, Nashville (Davidson County), 615-880-2001, www.nashvillefarmersmarket.org. ∏ ✈ @nashfarmmarket

West Nashville Farmers Market

The West Nashville Farmers Market springs up in Richland Park looking like a farm made to grow families. On a bright green lawn between the park's playground and a branch of the public library, the market is perfectly positioned to grow strong bodies and brains. A project of Good Food for Good People, this is one of several independent neighborhood markets in Nashville. All goods sold at the West Nashville Farmers Market are made or grown by their sellers. And when winter rolls around, the market doesn't just disappear; it simply moves to its indoor location.

46th Avenue North at Charlotte Pike, Nashville (Davidson County), 615-585-1294. ∏ ✈ @WestNashFrmsMkt

Dickson County Farmers Market

The Dickson County Farmers Market is a very popular place to be on a Saturday morning. That's no surprise, though, because not only will you find a great selection of produce, meats, breads, preserves, and crafts, but you'll also get to enjoy local musicians, cooking demonstrations, and agricultural education sessions. And it's not just a fun place for adults; the kids' corner gives them a place to have a great time while learning about farming and nature.

284 Cowan Road, Dickson (Dickson County), 615-446-2788, www.dicksonfarmersmarket.com. ∏

Franklin County Farmers Market

This small market feels like you're actually going out to the farm to buy your produce. Situated under the shade of big trees, the market is a place where you'll learn as much as you'll shop. While we were there, one farmer gave us a thorough tutorial about how he sets up his raised beds that were still producing beautiful summer squash in October.

Dinah Shore Boulevard, Winchester (Franklin County), 931-967-2741.

Giles County Farmers Market

Right beside the historic Giles County Courthouse in the Pulaski town square is a weekly market that feels like a street fair. At this busy market, you'll find everything from produce to crafts to baked goods to preserves to locally raised meats and everyone having a great time. And while you're in Pulaski, look out for turkeys. The city's art project has been a series of large painted Tom Turkeys that you'll find hidden in unexpected places all over town.

Courthouse Square, Pulaski (Giles County), 931-363-3789.

Lawrence County Farmers Market

In downtown Lawrenceburg, you can find a shady oasis in the summer. No, you can't sit and bask in cool water, but you can enjoy spending time with some of the friendliest and most knowledgeable vendors we've met who just can't seem to help giving extra attention and care to the products that they sell you. We bought blueberries, and when we left, we knew we had the best they had to offer. The vendors told us how we should keep them on the ride home (some of them actually made it there) and how we should store them and for how long (that wasn't a problem) and gave us ideas of what to do with them.

Mahr Street at Taylor Street, Lawrenceburg (Lawrence County), 931-762-5926.

Plowboy Produce Auction

This auction may just be one of the most fascinating and exciting things we saw on our travels. The auction brings two worlds together, providing a place for members of the local Amish community to sell their produce in bulk to those outside the Amish community. On one side of a long pavilion, Amish men driving horse-drawn wagons loaded with wares line up, waiting for their turn at the block. On the other side, trucks and vans wait to haul purchases to area stores and restaurants. Lots smaller than a wagonload are spread out under the pavilion for viewing. Some lots are even small enough to take home with you. We got quarts of beautiful black-berries, blueberries, and cherry tomatoes.

469 South Brace Road, Ethridge (Lawrence County), 931-829-1114.

Ardmore Farmers Market

Pop-up shelters line the edges of an open park at the Ardmore Community Center. Even with only a small group of vendors, the market is busy with people chatting in the park's shade and children playing. You'll find plenty of local produce here throughout the summer and early fall.

1646 Ardmore Highway, Ardmore (Lincoln County), 931-292-3110.

Marshall County Farmers Market

This market may be the most beautiful in the state. The market pavilion is across the road from Rock Creek Park and Walking Trail in Lewisburg. Scenic Rock Creek runs through the park, allowing a visit to the market to become a full day out. Under the pavilion, the farmers go beyond the traditional, offering more exotic produce such as tomatillos. Farmers also participate in demonstrations for area children so that they can learn more about agriculture.

300 Old Farmington Road, Lewisburg (Marshall County),
www.marshallfarmersmarket.com.

Clarksville Downtown Market

This large urban market hosts around thirty vendors from Tennessee and Kentucky with a full array of meat and dairy, fresh-baked goods, produce, honey, and preserved goods, along with arts and crafts. Music from local artists sets the atmosphere of the market, and there are educational and just plain fun activities for children. Surrounding the market is Clarks-ville's historic downtown, offering plenty of restaurants for breakfast be-fore or after the market or for just a drink to cool down in the summer heat. It's also just a block from the scenic Cumberland River, so great views are a given.

McGregor Park, Clarksville (Montgomery County), 931-645-7476,
www.clarksvilledowntownmarket.com. 🅕

The Clarksville Downtown Market is a popular destination on Saturday mornings.

Livingston-Overton County Farmers Market

"Oh, my, this market has been here for over twenty years," one vendor told us when we asked. "They just built this pavilion four or five years ago, though," another vendor shared. The history of the Livingston-Overton County Farmers Market was only a small part of the warm welcome we received when we stopped by in the fall. The market was slowing down, but there was still a great selection of end-of-season tomatoes, cucumbers, and peppers along with pumpkins, potatoes, sweet potatoes, and winter squash.

University Street at Spring Street, Livingston (Overton County).

Robertson County Farmers Market

This thriving farmers' market meets in Springfield and offers shoppers an array of goods from meat to soap to produce to honey. There was a lively crowd of regulars who seemed to be there to visit as much as to shop. Even toward the end of the season, there was a great selection of produce with beautiful tomatoes showing up throughout the market.

4635 U.S. Highway 41 North, Springfield (Robertson County). ◼

Murfreesboro Saturday Market

This market is perfect for strolling around as it wraps around one of only six remaining antebellum county courthouses in Tennessee. Unlike many markets, food is the sole focus around the square, where vendors offer produce, meat, and baked goods. Live music adds a festive feel, and it's easy to see that the market is as much a gathering place for its customers as it is a place to shop.

On the Square, Murfreesboro (Rutherford County). ◼

Rutherford County Farmers Market

This Rutherford County market takes place at the welcoming Lane Agri-Park Complex. The beautiful permanent facility here makes it easy for shoppers of any age to enjoy the bounty of the local harvest while staying cool and dry. In addition to being a great place for local vendors, the facility offers classrooms that are used to teach cooking and master gardening skills, among others. Outside, the Environmental Education Outdoor Garden is a wonderful learning place for children.

315 John R. Rice Boulevard, Murfreesboro (Rutherford County), 615-898-7710. ◼

Warren County Farmer's Market

Far from a simple country market, the Warren County Farmer's Market is a large and vibrant market under a beautiful permanent pavilion. You'll also find just about anything you could wish for, from the expected seasonal produce to exotics like Ping Tung eggplant and baby peacocks.

There's always something going on at the market. When we were there, the market was celebrating the top products of the fall season by hosting an apple baked goods contest and the Little Mr. and Miss Pumpkin contest that had cowboys, princesses, and monsters roaming the aisle. There's

The Warren County Farmer's Market is a hidden treasure.

a different attraction every week and plans to add even more, including local music.

East Colville Street at Market Street, McMinnville (Warren County), 931-473-8484, www.warrentnfarmersmarket.org.

Franklin Farmers Market

The Franklin Farmers Market is one of the few year-round farmers' markets in the state as well as one of the largest. Over the course of the year, over seventy family farms are part of the market. A visit yields a huge variety of meat, dairy, produce, baked goods, crafts, and anything else you might want. You'll also enjoy music and other entertainment, along with tasty samples.

But the market is about a lot more than what you see on a visit. Perhaps one of the most important projects of the market is the Growing Kids

Virtual Farmers' Markets

While one of our favorite parts of the farmers' market experience is the camaraderie of the people at the market, those crowds also mean competition for the available products. Occasionally, late risers can find themselves out of luck as things sell out. There are solutions for this, though. As farmers and farmers' markets are embracing technology, a new approach has emerged—the virtual farmers' market. Through websites like locallygrown.org, producers can list the type and amount of items they will have on market day. Users can order and arrange pickup online, ensuring they get what they want and leaving time to visit.

Garden. By working with the City of Franklin Parks, the farmers' market has planted a garden in the Park at Harlinsdale Farm specifically to teach kids about how plants grow; about how worms, bugs, and even toads are an important part of the garden; and about all the other roles that nature and farmers play to bring food to the table.

230 Franklin Road, Franklin (Williamson County), 615-916-1274, www.franklinfarmersmarket.com. **f**

CHOOSE-AND-CUT CHRISTMAS TREES

Blackjack Farm Christmas Trees

George Hofstetter has been growing Christmas trees for over thirty-five years. He was inspired to start the business by a field of Christmas trees he saw growing in East Tennessee. George talked as we huddled by the fire on a cold morning, but the afternoon would bring his favorite part of the day: "I love to see the families making a day of it. We have kids just running wild out there."

3331 U.S. Highway 231 North, Shelbyville (Bedford County), 931-437-2573, www.christmas-tree.com/real/tn/blackjack. **f**

Erin's Farm

Gary and Linda Hamm began their farm with a purchase of 75 acres that they have since expanded to 240 acres. They started by planting Christmas trees, and as long as they were doing the hard work of raising the trees, Linda decided to plant an acre of blueberries. Then Gary and Linda got involved with the USDA Forest Stewardship Program and began all sorts of new adventures. Now the farm is designed to protect the soil and provide a home for wildlife; their latest experiment is with shiitake mushrooms. All their other interests aside, though, what they look forward to every year is Christmas and the families that visit year after year for a hayride out to pick a tree followed by hot chocolate around a blazing fire.

5816 Hodges Road, Cunningham (Montgomery County), 931-980-3985, www.erinsfarm.com. ■

Santa's Place Christmas Tree Farm

"My father always wanted to do something involving Christmas when he retired," Eric Martin told us during our visit at Santa's Place. The result has been more than a simple field of Christmas trees. Ever the educator, Jerry Martin, after forty years of teaching, is always on the lookout for new information to share about Christmas; one path at the farm is lined with signs telling the story of Christmas around the world.

Jerry has also learned that, no matter how much you plan, sometimes your biggest successes come as a complete surprise. He built a small playground for visiting children, and to improve that area he had a load of gravel delivered to spread on the ground. "The kids ended up playing on the pile of gravel more than they did on the playground," Jerry said. He had the idea to top the mountain with a bale of straw studded with lollipops and call it Lollipop Mountain.

Jerry has also imported what he calls a genuine West Virginia hillbilly for the craft store at the farm. Eighty-year-old Brister makes everything from novelties to elegant decorations. When you're ready to head home, the Martins will prepare your tree for the trip and make sure it's securely on your car for the ride.

2175 Dunbar Road, Woodlawn (Montgomery County), 931-920-2744, www.santasplacetn.com.

Mark 4 Christmas Tree Farm

The four Galloways, Mary and Alan and their sons, Richard and Kyle, are one source of the name of their farm. Another meaning is from the Gospel of Mark, chapter four, which tells the parable of the sower of seeds. Of course, Alan associates the story with the physical work he and his family do at the farm, but he also thinks of the traditions started by the families who come every year to buy a tree to take into their own homes.

535 Kinniard Road, Cookville (Putnam County), 931-526-3398.

Rocky Point Tree Farm

When we visited Rocky Point, a young couple recently relocated from Indiana were talking to owner Richard Savage about how pleased they were to have found his farm so they could continue their Christmas tree tradition. While their tree was being bundled, their young daughter played around the miniature log house that serves as an office and windbreak. Richard smiled and waved as they drove away. Then we headed back to the cabin to warm up next to the tree covered in candy canes. Pointing to the tree, Richard told us, "That's what it's all about—the kids."

3214 Macedonia Cemetery Road, Cookeville (Putnam County), 931-526-2035.

Rhonda and Chris's Tree Land

Rhonda and Chris Leauber try to provide the best of both worlds. After you choose your tree, you can warm up with hot cocoa and fresh-baked cookies or sit around the campfire and toast marshmallows. But there is no piped-in music or lights. Instead, the Leaubers prefer to let nature stand out. The splashing of a small waterfall in the creek running along one side of the property and the birds singing in the trees all around it provide the music.

Chris manages the farm with the same kind of balance, a live-and-let-live approach to the wildlife with which he and his family share the area. On a tour of the farm, Chris does stop to pick off the occasional bagworm, a destructive pest that could decimate his crop if left unchecked. But he goes on to show us a row of trees that have been nearly completely stripped of their limbs and bark. "The deer come out here and rub themselves. It's worth sacrificing a few trees to be able to watch them," Chris says.

His other neighbors would seem to be more of a threat than they actually are. Two beavers have taken up residence in the lake next to the farm. The beavers have cut down some trees, but they don't bother the Christmas trees. "This is how we like it," Chris says. "We want people to come out here and stay as long as they want, and we give them as much or as little help as they want."

2054 Beech Log Road, Watertown (Wilson County), 615-237-9304, www.treelandtn.com.

WINERIES, BREWERIES, AND DISTILLERIES

Short Mountain Distillery
Billy Kaufman moved to rural Cannon County in 2001 and began work on a 300-acre organic farm with great concern for the protection of the water flowing through his land, not only for his sake but for that of the whole community. In 2010, he saw another opportunity to enrich the community and, with his brothers, began work on Short Mountain Distillery, a locally owned and family-funded business. Today, through the efforts of dedicated locals and the Kaufman family, the distillery produces small-batch sour mash moonshine, bourbon, and Tennessee whiskey from corn grown and stone-milled on Kaufman's farm.

119 Mountain Spirits Lane, Woodbury (Cannon County), 615-216-0830, www.shortmountaindistillery.com. ▯ ▯ @Short_Mountain

Beans Creek Winery
Beans Creek offers a good selection of wines made from Tennessee grapes, including tributes to the Bonnaroo Music Festival and the Swiss Heritage Festival in nearby Greutli-Laager. Other wines celebrate famous Tennessee walking horses that have been recognized at nearby Shelbyville's National Walking Horse Celebration. For entertainment, there's music on the grounds throughout the summer. To support local farmers, the winery now hosts a weekly farmers' market.

426 Ragsdale Road, Manchester (Coffee County), 931-723-2294, www.beanscreekwinery.com. ▯ ▯ @beanscreekwine

Water and Whiskey

Farmers look to the sky for rain to water the soil for their crops and pastures. Other Tennessee producers look to the earth for the water that provides their livelihood. Tennessee whiskey, by and large, is made from water flowing from limestone caves. Given the nationwide popularity of whiskey, most is no longer made using local corn, but the history is there, and the distilleries are worth a visit.

The Winery at Belle Meade Plantation

A visit to the Belle Meade winery is truly a step into the past. The winery is located at the heart of the thirty-acre site of Belle Meade Mansion, a beautiful example of antebellum architecture and decor. The plantation rose to fame through its stable of thoroughbreds in the early 1840s and remained a strong presence in the racing world until the early 1900s. Descendants of Belle Meade thoroughbreds can be found on racetracks throughout the country today.

Since opening in 2009, the winery celebrates the equestrian heritage of the plantation with the names and labels of its wines. A ticket for the tour of the house and grounds is not required to visit the winery, but your wine experience will be even better when you know the full story. Plus, all wine and ticket sales go toward preserving the mansion and grounds.

5025 Harding Pike, Nashville (Davidson County), 615-356-0501, www.thewineryatbellemeadeplantation.com. ⨍ 𝕏 @bmpwinery

Yazoo Brewing Company

We visited the Yazoo brewery on a Saturday afternoon, the only day of the week that the brewery offers tours. The taproom was filled with locals and tourists there to sample what Yazoo has on tap. While hops aren't grown in enough volume in Tennessee to support a brewery the size of Yazoo, the brewery makes sure that the flavors of Tennessee are strong in its beers. You'll find well-made India Pale Ale, American ale, Mexican-style beer, porter, stout, and hefeweizen, but you'll also find Gerst, a beer Yazoo formulated based on a recipe that vanished with Nashville's William Gerst Brewing Company when it closed in 1954 after sixty-one years in the city. You'll also find Sue, a smoked beer flavored with local woods. Snacks offered in the taproom include cheeses from Sweetwater Valley in East Tennessee and Kenny's Farmhouse Cheese from nearby Kentucky.

The brewery tour is minimally priced for adults and includes samples. Children are welcome at the brewery and can participate in the tour at no cost but, of course, also with no samples. Make sure to get there early as tours usually sell out at least an hour before the scheduled tour time.

910 Division Street, Nashville (Davidson County), 615-891-4649, www.yazoobrew.com. Tours given on Saturdays, 2:30 p.m. to 6:30 p.m. Dining $ 🅵 🅈 @YazooBrew

Highland Manor Winery

Opening in 1980, Highland Manor is the oldest winery in Tennessee. There are three acres of grapes grown on the grounds, but the majority of the grapes the winery uses come from other Tennessee vineyards. In addition to grapes, the grounds play host to several public events every year.

While wine is, of course, the focus of the vineyard, you won't want to miss out on some of the other treats the winery provides. Grape-seed oil pressed from Tennessee grapes is available, as well as a selection of artisan preserves, including pumpkin and muscadine butters.

2965 South York Highway, Jamestown (Fentress County), 931-879-9519, www.highlandmanorwinery.net.

Monteagle Winery

Since 2007, Monteagle Winery has been offering a delicious variety of Tennessee wines. While Denise Nunley and her staff grow grapes on over thirty acres, they can't always grow enough to meet their customers' demands. When they do have to use other grapes, though, they make sure to buy from local vineyards as much as possible. They would buy all locally, but there just aren't enough grapes being grown. As Denise told us, "Most people think that growing grapevines is going to be just like growing tomatoes. But it's not, and they quit once they realize that it takes so much hands-on work, and then they won't see any results from that work for years."

Still, Monteagle manages to produce a wide variety of wines that can please any palate, including the only sherry-style wine made by a Tennessee winery. And look for special events at the winery, where Denise and her staff mingle their wine with music and food for a day to remember.

847 West Main Street, Monteagle (Franklin County), 931-924-9400, www.monteaglewinery.com. **f**

Grinder's Switch Winery

The road that leads to Grinder's Switch will take you through rolling hills of rural countryside, and once you're there, you'll enjoy the peacefulness of the winery's wooded, parklike setting. Grinder's Switch is a relatively new winery, only in business since 2007, but its owners are dedicated to growing as much of their fruit as possible and are creating blends that speak of local flavors.

2119 Highway 50 West Loop, Centerville (Hickman County), 931-729-3690, www.gswinery.com. **f** 🐦 @GSWinery

Amber Falls Winery and Cellars

To reach the tasting room at Amber Falls, you drive up a winding road through lush vineyards to a beautiful building and deck that look like they belong more to a park than a winery. But as soon as you go down the stairs into the tasting room, you'll know exactly where you are. Wine barrels line the walls on one side of the room, while racks of bottles line the other side. While Amber Falls is most definitely a Tennessee winery, its South Louisiana roots show. Owners Judy and Tim Zaunbrecher made Tennessee their permanent home after Hurricane Katrina, and the winery hosts an annual Cajunfest each May and bottles a celebratory Cajun spiced wine.

794 Ridgetop Road, Hampshire (Lewis County), 931-285-0088, www.amberfallswinery.com. **f**

Keg Springs Winery

Since 2004, this small, family-operated winery has been offering Tennesseans a nice choice of wines made mostly from locally grown fruit. While the winery itself grows only Vidal grapes on its fifty-five acres, the Hamm family is dedicated to supporting local farms as much as possible. Keg Springs began when Brian and Becky Hamm decided to retire from corporate life; Brian's parents made Tennessee their home so that they could help make what had been a hobby into a business.

361 Keg Springs Road, Hampshire (Lewis County), 931-285-0589, www.kegsprings.com. **f**

Prichard's Distillery

Phil Prichard has a family history of making whiskey. Five generations ago, in 1822, Benjamin Prichard's will leaving his distillery equipment to his son is the only documentation of the last known legal distiller in the Prichard family until today. In 1997, Phil decided to reclaim the family legacy and opened his distillery in Kelso. His small-batch Tennessee whiskey uses locally grown white corn ground for him at the nearby historic Falls Mill and the finest spring water in Tennessee and is distilled in copper pots with the same techniques that Benjamin used all those years ago.

11 Kelso Smithland Road, Kelso (Lincoln County), 931-433-5454, www.prichardsdistillery.com. ◼

Red Barn Winery and Vineyards

Judy Clements takes the agricultural side of her winery very seriously. The vines have been growing here since 1994, and Judy and her husband, John, have been selling the fruits of their labor since 2000 from their repurposed tobacco barn. Judy puts their philosophy very simply: "If we don't grow it, I don't make wine with it."

That philosophy may limit the varieties they're able to offer, but Judy comes from an agricultural background that won't allow her to make any other choice. Judy is pragmatic about the choices other people make, too. "Anybody can go into a liquor store and buy a bottle of wine for the same prices we sell here, but coming to the winery is an experience, and we make sure that we give our visitors the best experience possible."

1805 Tanyard Road, Lafayette (Macon County), 615-688-6012, www.redbarnwinery.com. ◼

Beachaven Vineyards and Winery

The Cooke family has been making wine at Beachaven for over twenty-five years. Three-fourths of the grapes used in Beachaven's acclaimed wines are grown in Tennessee; three acres of grapes are grown on-site at the winery. While all of the Cookes' wines are good, their only purely Tennessee wine is their Seyval, and it's delicious. The winery also hosts Jazz on the Lawn, a periodic spring and summer music series that draws thousands of guests to relax at the vineyard.

1100 Dunlop Lane, Clarksville (Montgomery County), 931-645-8867, www.beachavenwinery.com. ◼ ◤ @BeachavenWine

Holly Ridge Winery and Vineyard

As lifelong Tennesseans, we still have trouble wrapping our heads around the concept of wineries in our state; it just seems too, well, too California. Holly Ridge Winery changed that for us. When we arrived, we were alone in the tasting room for a moment before Mrs. Wallin came in to tell us that her husband, Curtis, the winemaker, was on the way to guide us through a tasting. Sure enough, in just a moment, here he came, driving his tractor in from the vineyards. A winemaker on a tractor made it feel like home for us. Curtis is proud to grow all his own grapes. "They said I was too far north to be able to do that, but that was twelve years ago, and here we are," he says.

486 O'Neal Road, Livingston (Overton County), 931-823-8375, www.hollyridgewinery.com.

DelMonaco Winery

DelMonaco makes wine in a beautiful Italianate villa situated within the rolling hills of its vineyard. It's a perfect location for weddings, parties, and other events, and the winery hosts special events of its own throughout the year. But as beautiful as the winery is on the outside, its true colors show in the DelMonaco family's wine.

Their Espiritu de Oro (Golden Spirit) wine was made in partnership with Tennessee Technical University. Faculty and alumni helped to develop the blend of grapes grown in the winery's vineyard; students designed the label. But beyond working with the university to create a unique opportunity for learning, the winery donates a part of the proceeds of the sale of this wine to the university.

600 Lance Drive, Baxter (Putnam County), 931-858-1177, www.delmonacowinery.com. f

Long Hollow Winery

Stu Phillips is a good ol' boy from the hills—the foothills of the Canadian Rockies, that is. He calls the hills of Tennessee home now, though. He followed the siren song of country music to Nashville, where he's been a member of the Grand Ole Opry for over forty-five years. When Stu isn't on the Opry stage, he and his wife make wine just north of Nashville in a winery designed to resemble a monastery that Stu visited as a young man. And wine and music aren't separate loves for Stu. A combined tour of the winery and mini-concert by the Country Music Hall-of-Famer can be arranged.

665 Long Hollow Pike, Goodlettsville (Sumner County), 615-859-5559, www.longhollowwinery.com. **f**

Sumner Crest Winery

Sumner Crest opened in 1997, using fruit from the largest vineyard in Tennessee. After the death of one of the original founders in 2008, the winery began buying fruit from local growers, but the heritage of the winery is apparent as soon as you walk in the door. The tasting room is decorated with antiques collected over fifty years and restored to perfection. This winery is also invested in the future. Profits from Sumner Crest Pink Passion blush wine go to the Susan G. Komen Foundation to support breast cancer research.

5306 Old Highway 52, Portland (Sumner County), 615-325-4086, www.sumnercrestwinery.com. **f**

Arrington Vineyards

If you're looking to get your kicks in Tennessee, your best bet is Arrington Vineyards. Country music star Kix Brooks, of Brooks and Dunn fame, began his career in rough-and-tumble beer joints in Louisiana and ultimately co-founded this elegant winery near Nashville. Because of limited land, the winery is able to grow only about 15 percent of the grapes it needs, but it does make good use of what it grows. Arrington's Chambourcin grapes are made into one of the only port-style wines produced in Tennessee.

For Arrington, limited land is no obstacle for making good wine or great memories. By calling to schedule, visitors can take to the skies on a hot-air balloon ride or stay earthbound for a picnic on the grounds. Saturday evenings feature the Music in the Vine series of live jazz performances.

6211 Patton Road, Arrington (Williamson County), 615-395-0102, www.arringtonvineyards.com. **f** 🐦 @avwinery

STORES

Porter Road Butcher

James Peisker and Chris Carter both have impressive culinary résumés, but cooking is not their love. Instead, they look to raw ingredients. Their latest undertaking, Porter Road Butcher, is an old-style butcher shop with a few modern touches. They purchase whole animals from local farmers and break them down in the butcher shop, wasting as little of the animal as possible by using creativity and skill. If you're looking for a steak, some ribs, bacon, or something more exotic like tongue or fancy like pâté, these guys have you covered. And if you're looking for something that's not on the menu, just ask, because here, you never know.

501 Gallatin Avenue, Nashville (Davidson County), 615-650-4440, www.prbutcher.com. ⓕ ⋎ @PRButcher

The Turnip Truck

Grocery stores serve a simple purpose: to provide a source of food to a community. But the ways in which a store can serve that purpose are as varied as the people who operate them. The Turnip Truck uses its position in the community to provide local, sustainably produced goods in two Nashville locations.

When we visited the stores, we were amazed at how many local products we could find on the shelves and how competitively those products were priced. Along with locally produced goods, The Turnip Truck offers organic and all-natural foods and essentials.

970 Woodland Street, 615-650-3600; 321 12th Avenue South, 615-248-2000; Nashville (Davidson County), www.theturniptruck.com. ⓕ ⋎ @turniptrucknews

The Feed Mill

The Feed Mill labels itself "An Amish Country Market," and it most certainly is. In the store, you can find a wide variety of Amish products, from popcorn to baked goods. Other local products, like dairy and cheese, are available as well. When spring rolls around, the farmers' market outside takes off, rolling on into fall and finally ending with gourds and pumpkins. When the weather is warm, the dining options also grow. Every weekend, the grill is fired up for burgers and steaks for dinner, and a country breakfast is available on Saturdays. Diners can either eat inside, surrounded by photos and memorabilia showing the history of the area, or outside, next to the small creek that runs behind the store. But whether dining there, taking something home, or both, you're sure to enjoy your visit and leave satisfied.

7280 Nolensville Road, Nolensville (Williamson County), 615-776-4252, www.nolensvillefeedmill.com. Dining $ ▪️

DINING

Readyville Mill and Eatery
While the original water-powered wheel at the mill in Readyville is no longer running, the mill still produces organic wheat and corn products that are promptly taken to the kitchen next door to become your breakfast. The restaurant offers whole-wheat pancakes, corn cakes, and grits along with a special or two of the day. Local musicians provide entertainment in the dining room, and before or after your meal, you can visit the mill. If you want more than breakfast, Readyville Mill offers event facilities and prepacked bags of its freshly milled products.

5418 Murfreesboro Road, Readyville (Cannon County), 615-409-1405, www.readyvillemill.com. $ ▪️

Burger Up
Burger Up was born from a desire for what owner Miranda Whitcomb Pontes called "a community-driven restaurant." She wanted a space that would be welcoming to everyone from families to businessmen, and she hoped to host everything from special occasions to a fill-up after a soccer match.

Salvaged wooden planks make for furnishings that are at once elegant and whimsical, but the food makes Burger Up worth the trip, regardless of the decor. All the beef is from Triple L, a local ranch, and other ingredients are sourced locally if at all possible. The most popular burger, the Woodstock, is a complete celebration of local food, with the bacon and cheese from Tennessee and even Tennessee whiskey in the ketchup.

2901 12th Avenue South, Nashville (Davidson County), 615-279-3767; 401b Cool Springs Boulevard, Franklin (Williamson County), 615-503-9892, www.burger-up.com. $ ▪️ 🐦 @burgerup

Capitol Grille at the Hermitage Hotel

The Hermitage Hotel is only two blocks from the Tennessee State Capitol. Of more interest to diners at the Capitol Grille is that the farm supplying the produce is only five miles away. The biggest surprise is that chef Tyler Brown is also the farmer.

In 2008, the Hermitage began a program allowing guests to donate $2 per night spent at the hotel to the Land Trust for Tennessee. Those small donations have added up to over $157,000 so far. To show its gratitude, the land trust allowed the Capitol Grille to develop a garden at the sixty-six-acre Farm at Glen Leven, a property held by a single family since the Revolutionary War until it was donated to the trust in 2006. Tyler and his chef de cuisine, Cole Ellis, take full advantage of their bounty, using it for a menu filled with dishes inspired by southern classics. They are also looking to the future by expanding what they raise at the farm, including beef cattle.

231 6th Avenue North, Nashville (Davidson County), 615-345-7116, www.capitolgrillenashville.com. $$$ ⓕ 🐦 @capitolgrille

The Catbird Seat

Having local ingredients on the menu is important to many chefs. For chefs Josh Habiger and Erik Anderson, one thing is different—there is no menu. The Catbird Seat has room for only thirty-two diners, allowing Josh and Erik to create a different seven-course meal every night using exquisitely fresh ingredients. Reservations are hard to come by as the restaurant's approach has gained nationwide attention, but a visit is sure to be memorable.

1711 Division Street, Nashville (Davidson County), 615-810-8200, www.thecatbirdseatrestaurant.com. $$$$ ⓕ 🐦 @SHprojects

City House

Tandy Wilson was a college student when he took his first restaurant job as a dishwasher. Moving to a job in the kitchen, Tandy realized that he wanted to make cooking his future. After college, he went to Arizona for culinary school and then to California for a job. Next he traveled to Italy before returning home to Tennessee. Now chef/owner at City House, he is a supporter of the local food scene as well as of organizations like the Southern Foodways Alliance. His menu comprises a strong set of Italian classics and inspirations made with many local ingredients. Tandy's specialty and a must-try part of the menu are hand-cured meats and house-made cheeses like salami and mozzarella.

1222 4th Avenue North, Nashville (Davidson County), 615-736-5840,
www.cityhousenashville.com. $$ 🅵 🆈 @cityhouse

Flyte World Dining and Wine

When we first looked over the menu at Flyte, we were shocked at just how much local food we could pick out. For an elegant restaurant with a focus on wine, it would be easy to make local food less of a priority. But even in the months before the farmers' markets pick up, local products were represented in every course.

While we enjoyed our whole meal, the highlight for us was the soup and salad courses. We chose the "Flyte" option; instead of a large portion of one soup or salad, you can get a smaller portion of three of each. We were able to sample a nettle soup made with nettles from Appalachia and a butternut squash soup made with Nashville-area squash. One salad featured young ramps from the East, while another was complemented with locally produced goat cheese.

While Flyte might not be the best option for families with small children, for those looking for a romantic night of wine and great food, Flyte is a wonderful choice.

718 Division Street, Nashville (Davidson County), 615-255-6200,
www.flytenashville.com. $$$ 🅵 🆈 @FlyteNashville

Gabby's

Don't let the outside fool you. Gabby's may not look the part, but it's home to some of the best burgers in the state, all made from local grass-fed beef. Delicious food is important for any restaurant, but the people who work there make just as much of an impression. At Gabby's, it's obvious that the employees love what they're doing, and they make this diner-style restaurant a fun atmosphere for patrons of all ages. One thing to keep in mind is that Gabby's is small and very popular. If you choose to eat there at noon, be prepared for a line that often stretches out the door.

493 Humphreys Street, Nashville (Davidson County), 615-733-3119,
www.gabbysburgersandfries.com. $ 🅵

Miel Restaurant

Service is important to all restaurants, but Miel takes the concept of service beyond the table—beyond the restaurant, even. Chef David Maxwell and owner Seema Prasad run the Miel Community Kitchen, a once-a-week program designed to teach disadvantaged students of high school age culinary and job skills.

The dedication to the community is just as strong inside the restaurant, where local trout, pork, and beef go into the dishes served. As much of the produce used in the restaurant as possible comes from Miel's own farm, only ten minutes away on the Cumberland River.

343 53rd Avenue North, Nashville (Davidson County), 615-298-3663, www.mielrestaurant.com. $$$

Tayst

The highlight of the elegant decor at Tayst is the chalkboard just inside the front door listing the area farmers whose products are being used that day. Chef Jeremy Barlow turns those ingredients into exquisite dishes that please the palate and the eye. Not everything is a froufrou affair, however. The desserts are straightforward delights, with a superb crème brûlée and a bread pudding made from that southern classic, Krispy Kreme doughnuts.

2100 21st Avenue South, Nashville (Davidson County), 615-383-1953, www.taystrestaurant.com. $$$ @the_green_tayst

Tin Angel

Tin Angel is in a beautifully restored old building near Vanderbilt University, blending historic rusticity and modern flare. The menu is also an excellent mix of worldly and down-home. Lobster risotto and Spanish-style pork shanks stand alongside a burger made with local beef and bacon. Best of all is the "local bites platter," featuring prosciutto from East Tennessee's Benton's Country Hams, house-made sausage, cheese from a nearby dairy, and other local products.

3201 West End Avenue, Nashville (Davidson County), 615-298-3444, www.tinangel.net. $$ @TinAngelNash

The Wild Cow Vegetarian Restaurant

While we enjoy vegetables year-round, we're not vegetarians, and we were admittedly a little worried when we sat down to eat at The Wild Cow. We shouldn't have been and, in fact, were planning our return visit before we left the table from our first. The menu at The Wild Cow changes both seasonally and daily to reflect the freshest ingredients available. When we were there, we got to enjoy a black-eyed pea and local greens soup along with a salad with fresh lettuces, jicama, and toasted pumpkin seeds. With vegetarian and vegan options as well as a kids' menu that can please any palate, The Wild Cow is a great option for any family.

1896 Eastland Avenue, Nashville (Davidson County), 615-262-2717, www.thewildcow.com. $ 🄵 🅈 @TheWildCowVeg

Julia's Fine Foods

Julia Stubblebine began working with food in 1997, first offering a home delivery service and then catering. She opened Julia's Fine Foods in Sewanee in 2010 as a storefront for customers to pick up orders, but soon it turned into a busy restaurant. She works with local farmers to create a wide variety of dishes. One regular menu item is a locally raised free-range rotisserie chicken. Daily specials range from traditional southern barbecue to Korean barbecue tacos, both using locally raised meats and incorporating her Southern California flair with a big dose of southern charm. Of course that variety is natural at a restaurant with the motto "Always something different."

24 University Avenue, Sewanee (Franklin County), 931-598-5193, www.juliasfinefoods.com. $ 🄵 🅈 @JuliasFineFoods

Shenanigan's

Shenanigan's has been a gathering spot for the faculty and students of the University of the South since 1974. It offers the usual things: burgers, fries, and, best of all, a student discount. A few things separate Shenanigan's from the typical college joint, though. A cheeseburger made with ground beef and cheese from local producers and sandwiches on locally baked bread are favorite items. The highlight of the menu—for anyone with a sweet tooth, at least—is the selection of locally made desserts, including caramel and chocolate layer cakes.

12595 Sollace M. Freeman Highway, Sewanee (Franklin County), 931-598-5774, www.shenanigans-sewanee.com. $ 🄵 🅈 @ShenansSewanee

Joe Natural's Farm Store and Café

To really understand what's behind the fresh and delicious food on your plate, ask owners Deborah and Paul Schertz about their contributions from their own Roaring Creek Farm. In addition to what they raise, they buy as much meat, cheese, and produce as they can from other local farms. Local beef pot roast and chèvre with chipotle make the Santa Fe Kick sandwich a big hit at the restaurant. A tasty nod to local history and a clear sign that Joe Natural's is more than just a simple farmhouse restaurant is the Miss Mary Bobo, a roasted turkey sandwich with roasted fig, garlic chutney, and grilled onions. Miss Mary owned the boardinghouse where Tennessee whiskey giant Jack Daniel frequently ate lunch. Now Deborah and Paul are offering the same hospitality with their excellent farm-inspired and farm-sourced food.

4150 Old Hillsboro Road, Franklin (Williamson County), 615-595-2233; 209 10th Avenue South, Nashville (Davidson County), 615-345-6313, www.joenaturals.com. $ 🛇 🐦 @joenaturals

LODGING

Crocker Springs Bed and Breakfast

Bev Spangler takes well-justified pride in the restored 1880s farmhouse that is Crocker Springs Bed and Breakfast. But as beautiful as the house is, it's the outdoors that will really pull you to Crocker Springs. A rippling stream runs behind the house. Across a footbridge, you can explore a smokehouse, an outhouse, the old tenant house on the property, and the barn, where you can talk current events with Charlie the horse and Chewy the donkey. Or you can relax beside Bev's beautifully planned herb garden or on the deck overlooking the stream.

Unlike some bed and breakfast owners, the Spanglers love having families with children stay in their home, and they're happy to provide a place for the whole family to enjoy.

2382 Crocker Springs Road, Nashville (Davidson County), 615-876-8502, www.crockersprings.com. $$$

The Inn at Evins Mill

A historic gristmill may not seem a likely place to escape the daily grind, but The Inn at Evins Mill is an ideal relaxing getaway. The original mill was built on the site in 1824. In 1939, businessman and state senator Edgar Evins constructed the mill that currently stands on the property. The mill stopped running at the end of World War II, but a lodge that had been built on the property at the same time as the mill became a retreat for the Evins family. That log lodge has now been modernized with a conference area and additional guest rooms. Dining at the inn bridges modern comfort and Tennessee traditions with an elegant dining room and a large array of locally sourced ingredients. Whether you spend an afternoon strolling the grounds before dinner or a weekend in the inn, you can't help but leave Evins Mill more relaxed.

1535 Evins Mill Road, Smithville (DeKalb County), 800-383-2349, www.evinsmill.com. Lodging $$$$, dining $$$ 🅵 🆈 @evinsmill

The Bed and Breakfast at Falls Mill

Falls Mill was built on Factory Creek in 1873 as a cotton and wool factory and was later converted to a cotton gin and then to a woodworking shop; it now uses its waterwheel to power millstones that grind cornmeal, flour, and grits. A tour of the mill lets you see antiques like the millstones, the dog-powered butter churn, and all the handlooms and power looms of the weaving room. And if spending a day at the mill and exploring the beautiful grounds there just isn't enough, you don't have to go home. Falls Mill offers a cabin for overnight guests, well stocked with breakfast foods and goods from the mill.

134 Falls Mill Road, Belvidere (Franklin County), 931-469-7161, www.fallsmill.com. $$

Chestnut Hill Ranch Bed and Breakfast

When we visited Chestnut Hill, we were greeted with glasses of cold lemonade and the aroma of a delectable tomato sauce cooking slowly in the kitchen. Owner Cher Boisvert grew up in the hospitality industry in Canada, and when she and her husband purchased the ranch, opening their home as a bed and breakfast seemed to her to be a natural way to expand their social circle. Guests enjoy one of three themed rooms in the lovingly restored farmhouse as well as access to the fifty-three-acre working ranch, home to cattle, horses, donkeys, chickens, ducks, geese, and a pot-bellied pig. Cher provides her guests with gourmet evening meals, using the products of the farm as much as possible. The property also offers a pavilion and gazebo for weddings and other events, including educational opportunities for local schoolchildren.

3001 Browns Bend Road, Only (Hickman County), 931-729-0153, www.chestnuthillranch.com. $$

Enochs Farm House Inn

When you stay at Enochs, you don't just get a room in a farmhouse; you get an entire farmhouse along with access to a beautiful farm steeped in history. Little Blue Creek flows behind the house down to the millpond dug in the 1930s with an old road scoop and two stubborn mules. The gristmill helped the family through the Great Depression as it was used to grind corn for neighbors for a share of the output. It also provided electricity to the farm, using a salvaged generator. If you wish, Joyce Bullington, granddaughter of the farm's founder, or her grandson Nathan will gladly tell you the history of the mill and of the farm that's been in their family for over a century. They'll also sell you cornmeal ground at the mill along with homemade preserves and farm-fresh eggs.

3133 Little Blue Creek Road, McEwen (Humphreys County), 931-582-3218. $

Lairdland Farm Log Cabin Bed and Breakfast

Driving to Lairdland, we noticed a beautiful old cabin and commented that we wished we could stay there. Once we got to the office for the bed and breakfast, owner Jim Blackburn invited us to join him for a tour of the farm, and within minutes, we were touring that very cabin.

There are two cabins on the farm located across the street from pastured cows. The first is a restored 1835 beauty. The other is a "new" cabin built almost entirely from materials salvaged from other old buildings. Both cabins feature two bedrooms and two baths, but Jim has also restored privies (outhouses) to put behind each cabin. So far, no one has used them, but they're there just in case.

The "new" cabin is located behind the 1830s-era springhouse that provided water and cool storage for the farm. Both cabins offer access to a burbling creek and fifteen miles of trails for riding, biking, hiking, or just taking a stroll.

The main house is no longer part of the property, but it is now a Civil War museum that can be toured. Jim said, "A lot of times, people come out here to see the place, and they ask me, 'What is there to do out here?' I tell them about the creek and the trails, but when I tell them that most people just end up sitting on the porch swing and enjoying being out here, they always say, 'Yep. That sounds perfect.'"

3174 Blackburn Hollow Road, Cornersville (Marshall County), 931-363-9080. $$

Sequatchie Valley Bed and Breakfast Guest Ranch

Normally when you check into a bed and breakfast, you're graciously shown to your room, but things work a little differently at Sequatchie Valley. Here, when you check in, you get a horse. This will be a working vacation, but it also may be one of the best vacations of your life. You'll spend your days with your horse, riding trails, herding cattle, and learning to care for all of your horse's needs. You'll enjoy the great outdoors in the beautiful Sequatchie Valley every day that you're there.

1050 Ray Hixson Road, Dunlap (Sequatchie County), 423-554-4677, www.tnhorsevacation.com. $$$$ [facebook]

Cherry Hill Farms Bed and Breakfast

Passing through each of the gates that separate the pastures at Cherry Hill, you realize you are on a working farm. The winding drive leads up the hill to a beautiful home that can best be described as a chateau. Owner Christi Gamble originally lived in California and Nevada where horses were always part of her life and business. "I'm just a cowgirl!" she laughs. When her husband, Mike, wanted to return to his roots in Middle Tennessee, she agreed and suggested that they open their home as a bed and breakfast and horse farm.

The farm has become an elegant retreat for adults, with gourmet dinners and spa treatments available on request. And yet it's a homey place where repeat guests often bring other family members or friends for a stay where they can all be together to relax.

Cherry Hill isn't just a bed and breakfast for people, though. Equine guests are as welcome as their human counterparts and can be just as pampered while enjoying the pastures and rolling hills of the farm.

1104 Richmond Road, Watertown (Wilson County), 615-237-0471, www.cherryhillfarmsbedandbreakfast.com. $$–$$$

SPECIAL EVENTS AND ATTRACTIONS

The Farm at Hollow Springs

Having reached their seventieth wedding anniversary, Lisa Trail's parents have started taking life a little easier. Lisa and her husband, Tim Tipps, now own the land across the road from the house where Lisa grew up and are planning to farm it in the future. A visit to Hollow Springs will make you feel like you're spending an old-fashioned day in the country. The maze here is sorghum and was planned and planted with the help of the agriculture department at Middle Tennessee State University. Once the maze is done being used, a neighboring farmer will harvest the sorghum to use as winter feed for his animals, letting nothing go to waste. But the sorghum maze isn't the only maze you'll find here. Your mind will be challenged to figure out the right-turn-only hay maze, the numbers maze, and an interwoven string maze that will test your agility as well.

9190 Hollow Springs Road, Bradyville (Cannon County), 615-848-2822. �

Nashville Zoo at Grassmere

Most zoos feature an area to simulate a farm, but not all zoos once were farms. The Nashville Zoo stands on the former Grassmere Plantation. The Croft sisters donated their home to the Children's Museum of Nashville with the stipulation that the land always be used for nature education. The original 1810 plantation house still stands on the zoo's 188 acres. While the cropland of the plantation is part of the zoo now, the garden that sustained the family is maintained by the Master Gardeners of Nashville, who offer classes there.

3777 Nolensville Pike, Nashville (Davidson County), 615-833-1534, www.nashvillezoo.org. 🗎 🐦 @NashvilleZoo

Purity Dairy Tour

Purity Dairies began in 1925 with a single truck. Along the way, the company survived the Great Depression and World War II by innovating with technology and marketing. Today, the company is under the Dean Foods umbrella, but the people and plant are still Nashville, and the milk is still Tennessee. A tour includes a review of the company's history, a look at the butter-making equipment, and ice cream samples.

360 Murfreesboro Road, Nashville (Davidson County), 615-244-1970, www.puritydairies.com. 🗎

State Support for Agriculture

The Tennessee Department of Agriculture offers several programs to support agriculture and agritourism in the state. The Tennessee Agricultural Enhancement Program provides training and cost-sharing funds for all sorts of projects on farms. If you visit Dan and Debbie Eiser at Blueberries on the Buffalo, you may not notice it, but their new irrigation system helps them protect their blueberry bushes through droughts and freezes. The addition is more obvious at Walnut Ridge Llama Farm. Jerry and Carolyn Ayers were able to add a pavilion and store. The pavilion gives them more space for guests during their tours and fall activities, and the store gives Carolyn a way to directly sell the yarn and woven products she makes from fibers from their llamas.

The Pick Tennessee Products (PTP) program collects information from farms about their products and presents it on its very useful website, www.picktnproducts.org. Consumers across the state can find local produce through the site. The program also partners with the Tennessee Grocers and Convenience Store Association to put on a trade show where participants in PTP offer booths that will be seen by buyers from stores across the state.

Tennessee Agricultural Museum at Ellington Agricultural Center

A onetime horse barn now serves as home to over 200 years of Tennessee farm history. The grounds surrounding the museum feature typical buildings from nineteenth-century farm life. School groups and families can see demos of rural activities like quilting, butter churning, and storytelling at the Historic Rural Life Festival in May and the Music and Molasses Arts and Crafts Festival in October.

440 Hogan Road, Nashville (Davidson County), 615-837-5197, www.tnagmuseum.org.

A Toast to Tennessee Wine Festival

This celebration of Tennessee wineries offers festivalgoers a chance to sample Tennessee wines ranging from sparkling to dry table wines to sweet dessert wines. Of course, no wine tasting would be complete without food to pair with the wines, so Tennessee gourmet food producers are on hand with their specialty items made with local ingredients. The festival also offers an educational component with seminars on topics that have included food and wine pairings, cooking with wine, wine storage, grape varieties, and more. And with the event taking place at Nashville Shores, free lake cruises are also offered hourly to allow a relaxing break from the fun.

4001 Bell Road, Hermitage (Davidson County), 615-758-3478, www.atoasttotennessee.com. Held in May. ⓕ ⓨ @Toast2TNWine

Allardt Pumpkin Festival and Weigh-Off

The water tower for the town of Allardt is painted pumpkin orange, but even it seems to pale in comparison to the enormous pumpkins that are judged by the professionals to determine which local farmer has raised the biggest of them all. Since 1991, this festival has brought crowds to view the weighing and to listen to the secrets that these farmers bring to the festival along with their weighty produce.

Of course, looking at pumpkins is likely to get you thinking about pumpkin goodies. Don't worry; this festival has you covered. Local vendors and groups offer pumpkin concessions beyond counting. We found pumpkin pies, pumpkin rolls, pumpkin cinnamon rolls, pumpkin bread, pumpkin stack cakes, pumpkin cookies, pumpkin preserves, pumpkin butter, and even pumpkin ice cream topping. And I'm sure we missed some.

With a huge open car show, carnival rides, local crafts, and a costume contest for children of all ages, this festival has something for everyone to enjoy. And yes, there are plenty of farmers selling pumpkins of every description that might find a home on your porch or table.

Downtown Allardt (Fentress County), 931-879-7125, www.allardtpumpkinfestival.com. Held in October. ⓕ ⓨ @giantpumpkin

Poke Sallet Festival

"We're the eleventh oldest town in Tennessee. We were an important river town, but then the railroad went through Crossville and our town just dried up. All we have left now is history." So said the docent at the Jackson County Historic Society in Gainesboro. But come spring, they also have

At the Allardt Pumpkin Festival and Weigh-Off, you're in the big time.

a festival celebrating poke sallet, for many an obnoxious weed but once an important food source. The festival includes a carnival, arts and crafts, and, best of all, a chance to sample poke sallet.

Downtown Gainesboro (Jackson County), 931-268-0971, www.pokesalletfest.com. Held in May.

Goats, Music, and More Festival
The annual Goats, Music, and More Festival honors the fainting goat and the Boer goat with show competitions taking place all weekend for these breeds and others. The goats are judged for breed characteristics including coloration, size, and shape, but walk through the holding tent, and you'll find all of them beautiful and see true loving care from the farmers who get them ready for their big moments.

You'll also find arts and crafts, food vendors, music of all varieties, and barbecue cook-off competitions. You might even find goat barbecue if you look for it. Rock Creek Park is a lovely setting for the festival, offering plenty of picnic and open space for cooling off and relaxing during the day or for enjoying the free concerts at night.

505 North Ellington Parkway, Lewisburg (Marshall County), 931-359-1544, www.goatsmusicandmore.com. Held in October.

Mule Day Celebration

The Mule Day Celebration stock barn is a hotbed of intrigue as mules waiting for competition and judging bray gossip to their friends at the other end of the barn. During our visit, one poor mule, distracted by an ear scratching from a child, had her lunch stolen when a neighbor slipped his head through the bars of the pen.

In the nearby arena, mule teams compete, pulling wagons through an obstacle course while they are judged for teamwork and precision. Mule-powered demos are given to show how logs were loaded onto wagons in the days before diesel engines. The festival includes a parade, a beauty pageant (of the people variety), and a midway, where crafts and food are sold. In our opinion, though, the best food option is the beans and corn bread supper fund-raiser for the Maury County Senior Center.

1018 Maury County Park Drive, Columbia (Maury County), www.muleday.org. Held in April.

Riverview Mounds Century Farm

Philosopher-farmer Chris Rinehart grew up on the farm that has been in his family for nearly 180 years. He left the farm to join the navy and see the world; during his time in the military, he met and married Scarlett Mulligan. While they were both attending college in Oregon, they got their first taste of agritourism.

When Chris and his brother, Steve, inherited the farm, they knew they wanted to keep it in the family, but what to do with it? The answer became clear when Chris and Scarlett visited a pumpkin patch with their son—agritourism.

Now they open the farm to visitors every fall to provide a fun educational experience. Along with a pumpkin patch, corn maze, playground, and a host of animals, they show and teach interested visitors about the two Mississippian Indian mounds that give the farm its name.

"People have been living on this bluff for thousands of years. That's always been an amazing thing to think about for me," Chris told us. "And I want people who come here to feel that. We always say, 'This land is your land; this land is my land.' That's really what it's all about."

1715 Boyd Rinehart Road, Clarksville (Montgomery County), 931-624-1095, www.riverviewmoundsfarm.com. ⨍

Clouds of sweet steam fill the air at Muddy Pond Sorghum Mill.

Muddy Pond Sorghum Mill

It's not a complete spa treatment, but a visit on cooking days to Muddy Pond Sorghum Mill is the sweetest steam bath you'll ever take. The smell of boiling cane juice hits you as soon as you step into the parking lot and only serves to draw you in closer. The Guenther family has been making sorghum syrup for fifty years, and while the machinery may be more modern today than it was in 1960, the knowledge required to produce it is far older.

At Muddy Pond, you can buy sorghum, but even better, you can view the sorghum-making process from the very beginning. Outside, you can see the sorghum cane crushed for juice by a team of horses walking in circles to turn a press. Inside, the cauldron is no longer used; instead, a gravity-fed evaporator is used to thicken the juice into syrup. Still, the sorghum maker must watch carefully and tap years of expertise to decide when the syrup is ready to be bottled to go home with you.

4064 Muddy Pond Road, Monterey (Overton County), 931-445-3509, www.muddypondsorghum.com. 🔘

Sorghum Syrup, Sorghum Molasses, or Just Plain Molasses?

Technically, molasses refers to the thick syrup that's a byproduct of making crystallized sugar from sugar cane. It's common in the Caribbean and south Louisiana, where Steen's is one of the most popular producers of both cane syrup and molasses.

But many Tennesseans grew up eating what their parents and grandparents called sorghum molasses. While there's no real difference between sorghum molasses and sorghum syrup other than the name, many people complain that the sorghum syrup of today is milder and sweeter than the molasses they grew up with.

They do have a valid point. Much of the sorghum grown today is milder in flavor so that it will appeal to a larger audience. It's what you'll find on supermarket shelves and from most large-quantity producers. But it's not all that's out there. Small farmers and producers are growing heirloom varieties that keep that almost astringent flavor that made sorghum so very different from any other sweetener.

So what can you do with each one? Most varieties marketed as sorghum syrup have a slightly different, stronger flavor than honey, but they are perfect to use as a sweetener in granola, oatmeal, or almost anywhere you would use honey. The old-school-style sorghum molasses or sugar cane molasses makes the perfect topper for a hot biscuit once a knob of butter has been cut into the syrup. Either one is also responsible for an inimitable depth of flavor in gingerbread, molasses cookies, and barbecue sauce.

The Sorghum Mill

The Muddy Pond community is known for sorghum making, and while you'll find it from the local producers year-round on local store shelves, you'll get to see it growing in the fields and being cooked down into thick syrup only in September and October. While The Sorghum Mill is less well known outside the community, the Mazelins who operate it are justifiably proud of the syrup they produce, a slightly darker, thicker syrup that is harder to find than most. You can also watch them making syrup at the annual Ketner's Mill Fair in East Tennessee, where the Mazelins demonstrate the complete process from crushing the stalks to cooking the syrup down to the perfect consistency. And either place you meet them, don't hesitate to ask for a sample.

4570 Muddy Pond Road, Monterey (Overton County), 931-863-3859.

Armstrong Pie Company

Let your nose lead you to the Armstrong Pie Company in Linden for a tour that your mouth will thank you for. The company began in 1946 in the nearby town of Hohenwald, and very little has changed over the years. This is a small pie company. You'll get to watch the process that four employees follow to produce over 4,000 fried pies and miniature pecan pies every day with fillings made from as much local produce as possible. Local favorites include chocolate, Bavarian cream, coconut, and chess. It's not a long tour, but your stomach will be grumbling by the end. Not to worry, though; the folks at Armstrong won't let you leave hungry. Taking this tour is the only way you can get to enjoy a still-warm unwrapped pie, what has to be one of the best treats in Tennessee. And since one pie is never enough, be sure to take home a box of assorted flavors to eat later on.

106 South Mill Street, Linden (Perry County), 931-589-5009, www.armstrongpie.com. f

Homeplace 1850

From the moment you enter the interpretive center at the Homeplace, you'll realize you've stepped into a special place. While it's tempting to rush through the center and get to the fun outside, it's worth taking the time to study the displays of farm life during each season to get a real picture of what daily life at this working farm is like.

The crops and livestock at the farm are historic varieties that would have been there in the mid-nineteenth century. A visit can be a hands-on experience. If plowing is being done or repairs are being made to fences or buildings, you can pitch in. You'll also see household chores including cooking, spinning and weaving, sewing, and food preservation. If you want to see particular aspects of farm life like curing hams or plowing, be sure to check the calendar to get specific dates.

Located in Land Between the Lakes National Recreation Area, a visit to the Homeplace offers an opportunity for a great day in a beautiful piece of nature. And while the park address is in Kentucky, the Homeplace is located in the Tennessee portion of the park.

Pryor Hollow (Stewart County), 270-924-2020,
www.lbl.org/HPGate.html.

Middle Tennessee Strawberry Festival

The whole town of Portland gets involved in the strawberry festival—even the fast food restaurants along the square offer strawberry desserts and drinks for thirsty festival goers. You'll find a good variety of strawberry-themed concessions like strawberry ice cream and strawberry milk at this festival, but strawberry shortcake is by far the star here. And if you haven't gotten your fill of strawberries before you leave, there are plenty of local farms selling strawberries for you to take home.

Downtown Portland (Sumner County), 615-325-9032,
www.middletnstrawberryfestival.com. Held in May.

Fiddler's Grove Historical Village

Fiddler's Grove is located on the grounds of the James E. Ward Agricultural Center and exists to preserve the buildings, artifacts, and oral and written histories of Wilson County. This site consists of over fifty buildings and exhibits that showcase farm life from 1795 to 1950. While the village is open every Saturday, keep an eye on the village's calendar for second Saturday themed events that provide more in-depth looks at different aspects of village life and end with a dinner of soups, corn bread, and peach cobbler while local bluegrass performers play.

945 East Baddour Parkway, Lebanon (Wilson County), 615-443-2626,
www.fiddlersgrove.org.

Garlicky Kale

Melanie Cochran and the rest of the folks at The Wild Cow use typical vegetarian ingredients like tofu and seitan to make delicious dishes. Their real strong point, however, is allowing vegetables to stand on their own. This dish takes nutrient-rich kale and livens it up with garlic and jalapeño for a simple but excellent side dish.

SERVES 2

1	teaspoon olive oil
1	teaspoon minced garlic
1	teaspoon minced jalapeño
2	handfuls farm-fresh de-stemmed kale
½	teaspoon soy sauce or Bragg's Liquid Amino's

TO SERVE:
Cooked beans and rice

Heat the oil in a large skillet over medium heat.

Add the garlic and jalapeño and cook, stirring constantly, for 30 seconds.

Add the kale to the skillet.

Immediately add 2 tablespoons of water, cover the pan, and turn off the heat.

Allow the kale to steam, covered, for 1 to 3 minutes, depending on how soft you want the kale to become.

Uncover the pan and add the soy sauce or Bragg's, using tongs to toss and remove from the pan.

Serve alone or with beans and rice.

Green Pea, Watermelon, and Wheat Berry Salad

Wynn Jeter, executive chef at Joe Natural's, uses farm-fresh ingredients to create a menu of scrumptious sandwiches, soups, and salads. This recipe may seem unusual, but it shows that all you need for a tasty dish is imagination and quality products.

SERVES 4

1	cup uncooked wheat berries
1	cup shelled green peas
¼	teaspoon salt
¼	teaspoon freshly ground black pepper
1½	cups cubed seedless watermelon
1	cup chopped flat parsley
	Zest of 1 lemon
½	cup shaved Parmigiano-Reggiano cheese

Place the wheat berries in a large saucepan and cover them with water by 1 to 2 inches. Bring the water to a boil over medium-high heat, reduce the heat to low, and simmer, uncovered, until the wheat berries are tender, about 25 minutes.

Add the green peas to the saucepan with the wheat berries and continue cooking for 2 minutes.

Drain the wheat berry and pea mixture and rinse with cold water.

In a large salad bowl, combine the wheat berry and pea mixture with the salt and pepper, stirring to combine thoroughly.

Add the watermelon and parsley and toss to combine.

Sprinkle with lemon zest and Parmigiano-Reggiano.

Mule Day Chess Pie

One of the most popular stops at the Mule Day Celebration is the Maury County Senior Center, where you can enjoy hearty servings of white beans and corn bread. But you can't leave without a slice of the chess pie that the senior center is famous for. When you make it at home, this classic southern recipe is better than ever with farm-fresh eggs.

SERVES 8

1 ½	cups sugar
3	tablespoons evaporated milk
1	tablespoon white vinegar
1 ½	tablespoons cornmeal
1	stick unsalted butter
1	teaspoon vanilla
3	eggs
1	pie shell

Preheat the oven to 350 degrees.

Using a stand mixer, combine the sugar, milk, vinegar, cornmeal, butter, vanilla, and eggs until well blended.

Pour the mixture into the pie shell. Smooth the top of the pie with a spatula.

Bake for 45 to 55 minutes or until the top of the pie is golden brown.

Allow the pie to rest for at least 30 minutes before cutting and serving.

Tomato Gravy

Chefs Tyler Brown and Cole Ellis of the Capitol Hotel in Nashville have tremendous talent and a great kitchen. Another advantage they have is their own garden, giving them a private source for the freshest produce possible. One menu feature they make with their bounty is a southern tradition, tomato gravy. The rich, thick gravy can be used for many dishes, but one of the best uses is one of the simplest—over biscuits. Serve it hot in a bowl, or for a more elegant touch, fill a baking dish with the gravy then carefully add poached eggs. Either way, serve it with fresh hot biscuits and a big spoon for scooping.

SERVES 8

4	pounds fresh tomatoes (about 15 medium tomatoes) or 3 quarts home-canned tomatoes or 3 (28-ounce) cans peeled diced tomatoes
2	tablespoons bacon grease or 4 strips bacon, chopped
2	red bell peppers
2	green bell peppers
2	shallots
1	clove garlic
¼	cup white wine
	Salt and freshly ground black pepper to taste

TO PEEL FRESH TOMATOES:

Skip this step if using canned tomatoes.

Fill a large saucepan halfway with water and bring the water to a boil over high heat. Fill a large heatproof bowl halfway with ice water. Cut a small "X" on the bottom of each tomato. Using a slotted spoon, carefully lower the tomatoes singly or in small batches into the boiling water.

After 20–30 seconds, use the slotted spoon to transfer the tomatoes from the boiling water to the ice water. Allow the tomatoes to cool for 1 minute. They can now be easily peeled starting from the small "X" you cut.

If you wish to seed the tomatoes for the smoothest texture, once they have been peeled, place a fine mesh sieve in a bowl. Have another bowl available. Hold a tomato over the sieve and carefully cut it in half from side to side. Using your fingers, reach into the cavities of the tomato, scooping the seeds out into the sieve. After the seeds are removed, place the tomato in the other bowl. Repeat until all tomatoes have been seeded.

Discard the tomato seeds. Add any tomato juice that fell through the sieve to the bowl with the tomatoes.

TO MAKE THE GRAVY:

If you don't have bacon grease available, cook the bacon in a large skillet on medium-low heat until all the fat is rendered out, being careful not to burn it.

Transfer the bacon to a paper-towel lined plate to drain, leaving as much of the drippings in the pan as possible. Remove the skillet from the heat. The bacon can be stored for other uses once it has cooled.

Seed the peppers, discarding the seeds. Finely dice the peppers and mince the shallot and garlic by hand, or roughly chop the vegetables and pulse them in a food processor until they are finely minced.

If you had bacon grease, add it to a large skillet over medium heat. If you didn't, put the skillet with the bacon grease over medium heat.

Add the pepper, shallot, and garlic mixture to the skillet and cook until translucent and soft, about 8 to 10 minutes.

Add the tomatoes and wine to the skillet. If using canned tomatoes, add the liquid as well. Increase the heat to medium-high. Stir the tomatoes frequently until the consistency of gravy is reached. This will take about two hours depending on the amount of liquid in the tomatoes.

Season to taste with salt and pepper.

Scott

Campbell

Claiborne
• Tazewell

Hancock

Hawkins

Blountville Bristol

Kingsport
Sullivan

Shady Valle

Morgan

Anderson

Union
Rutledge •

Grainger

Russelville •

Norris •

Clinton •

Heiskell •

Corryton •

Knox

Oak Ridge •

Jefferson City

Jefferson

Knoxville •

Dandridge •

Gray
Johnson City •

Rogersville •

Bulls Gap •

Chuckey •

Greeneville •

Greene

Mountain City

Washington

Johnson Cty

Jonesborough

Carter

Limestone •

Unicoi

Johnson

Elizabethton

Erwin •

Unicoi •

Unicoi

Cumberland

Crossville •

Harriman •

Roane

Parrottsville •

Newport •

Cocke

Cosby •

Flag Pond

Pikeville •

Grandview •

Loudon •

Sevierville •

Philadelphia

Loudon

Maryville •

Sevier

Pigeon Forge •

Spring City •

Greenback •

Grandview •

Meigs

Blount

Gatlinburg •

Bledsoe

Rhea

Dayton •

Sweetwater •

Athens •

Madisonville •

McMinn

Tellico Plains •

Whitwell •

Charleston •

Monroe

Sequatchie •

Signal Mountain •

Cleveland •

Delano •

Reliance •

Marion

Hamilton

Benton •

Chattanooga •

Bradley

Polk

South Pittsburg •

East Tennessee

East Tennessee is the most geographically diverse of the three Grand Divisions of Tennessee. In the east, the Blue Ridge Mountains form the border with North Carolina. Treasures are hidden in small valleys and on the sides of mountains all through the region. In the center of the region, the wide Tennessee Valley is a rich crop-growing area. Wineries dot the sloping sides of the valley, and farms abound on the valley floor. Knoxville and Chattanooga grew on the banks of the Tennessee River and now serve as centers of commerce in the region. To the west, the Cumberland Plateau rises steeply from the valley. Now the plateau is home to farms and orchards, but in earlier history, it was sparsely populated due to the daunting barrier it formed.

FARMS

Mountain Meadows Farm

You'll see Ernie Meadows and his family selling their beautiful produce at most farmers' markets in the north corner of East Tennessee. Their small family farm is well known in the area for a great variety of fruits, berries, and vegetables. You'll find produce here that you won't see anywhere else, partially because Ernie likes to surprise his CSA customers with something special in their baskets with each delivery. There's usually some left over for the farmers' markets, too.

A lot of experience is behind the family's success. While the Meadowses have been marketing their produce in the Knoxville–Oak Ridge area only since 2001, they've been farming for over forty years. All of their produce is raised sustainably, with a care for the soil to maintain a healthy growing environment. Also, while you'll see them at several farmers' markets, none of their produce travels more than twenty-five miles from the farm.

284 Bridges Road, Heiskell (Anderson County), 865-494-9709, www.mountainmeadowsfarmtn.com. f

Colvin Family Farm

The Colvin family has a long history of farming in Tennessee, but current patriarch Steve left farming for nearly twenty years for other work before he and his wife, Val, realized their dream of buying a farm. The early years were spent reclaiming the land, but soon they were growing enough produce to run a successful CSA. After a few years, they needed more land to keep up with demand. Now the Colvins have added nearly sixty acres and are in position to grow more goodness than ever and to welcome visitors. Check their website for special events, or call to schedule a visit.

1045 JB Swafford Road, Spring City (Bledsoe County), 423-802-0462, www.colvinfamilyfarm.com. f y @Colvin_CSA_Farm

Circle A Farm

From a holler—a low, sheltered area in the mountains—Circle A Farm helps to support a worldwide ministry with the proceeds of its operations. Morris and Darlenia Anderson offer fun and educational activities for children and adults alike. While exploring the farm's thirty-seven acres, visitors can help tend to the gardens and the livestock, feed the fish in the lake at the center of the farm, or walk around the lake and get a lesson about the natural world surrounding the farm. If you want an even more immersive experience, overnight stays on the farm are available for any size group.

2349 Jericho Road, Maryville (Blount County), 865-984-6982, www.circleafarm.com. f

The Greenway Table

Eat well. Move often. Laugh together. These three simple statements say everything about the motivation behind the creation of The Greenway Table in Cleveland. This urban farm exists to educate city residents, area students, and local farmers about sustainable farming and the importance of healthy food choices. A CSA option allows locals to support this non-

Urban Farms

People talk a lot about consolidation in farming and problems for small family farms, but as this book shows, things are looking up. One of the strongest forces behind the revitalization of small farms—and the creation of new ones—comes from the city. As farmers have begun to see the increasing demand for goods at farmers' markets, they are taking advantage of the situation and selling their products directly to the public at retail prices rather than getting wholesale from a reseller.

Another interesting movement in farming is also tied to the city—or, rather, is actually surrounded by the city. Urban farms are popping up across the state. In Memphis, Grow Memphis has the mission of eliminating hunger and improving health. One way the organization is doing this is by turning vacant lots into organic gardens. Grow Memphis provides start-up funds and expertise. Once the gardens are established, the produce is distributed among the community, and some is sold to cover expenses.

Gardens on private property are also becoming big business in cities. Companies offering garden services are popping up everywhere. These services range from consulting on how and where to build your garden to outright construction of your own year-round mini-farm.

profit effort financially while benefiting greatly from the lessons it produces. While the farm is always open to visitors, keep an eye on its Facebook page and website for special events that will let you and your family get your hands dirty and pitch in while having fun.

315 20th Street Northeast, Cleveland (Bradley County), 423-790-0660, www.thegreenwaytable.org.

Beck Mountain

The Gentrys raise beef cattle, corn, and hay on their Elizabethton farm. They first opened the farm to the public in 2004 with a corn maze in a valley and one of the most beautiful hayrides around. Instead of just transporting you from a barn to the maze or a pumpkin patch, this hayride is designed to take you around the mountain so that you can enjoy the fall colors and all the great views. The Gentrys decided in 2006 to open an entertainment barn, where they offer their Fourth of July celebration with games, music, and more, all followed by a fireworks display over the mountain.

144 Webb Hollow Loop, Elizabethton (Carter County), 423-543-1045, www.beckmountaincornmaze.com. **f**

Mountain Hollow Farm

Beth Bohnert spent over a year researching goats before she purchased her herd of cashmere goats. But being a new goat farmer is a lot like being a new mother. "I had read so much that I thought I was ready for anything, but when things really started happening, I had no idea what I had gotten myself into," she recalls.

Her goats have done very well, and she also raises angora rabbits and llamas and keeps ducks for pest control. The goats are hardy, and Beth works to breed only good-tempered goats. Their soft wool is sent to a mill in Connecticut, where it becomes luxurious yarn that Beth sells in her shop on-site at the farm.

This shop is a little piece of local history. By all accounts, it's been there since at least the early 1930s, but some estimates put it there much earlier as a Native American trading post. Today, its walls are lined with colorful yarn, gourmet foods, and loose-leaf teas, and it's become a gathering place for knitting and crocheting enthusiasts from near and far.

553 Vancel Road, Tazewell (Claiborne County), 423-869-8927, www.mtnhollow.com. **f y** @MHFCashmere

Hicks Family Farm

Georgia Hicks started growing pumpkins and winter squash in her garden in the late 1970s while her husband tended to the rest of their cattle and tobacco farm. When people began asking to buy her pumpkins, she started selling them from the front yard, and a new business for the farm was born. Today, her grandson Brad raises over twenty acres of pumpkins, fall squash, and gourds, including some of the biggest pumpkins we've seen in a field. In 2007, Georgia and her family decided to take another venture with their farm and opened a corn maze that's growing as well as Georgia's pumpkin business did.

380 Wilton Springs Road, Newport (Cocke County), 865-322-2377, www.hicksfamilyfarm.com. **f**

The Pumpkin Patch at Neas Farm

Kevin Neas is part of the fifth generation of his family on this century farm, where he raises beef cattle along with traditional burley tobacco and wheat. While farming is the business, his wife, Darbi, is just as concerned about preserving the heritage of the farm. She made sure that the house and original outbuildings here are on the historic register and added the farm to the Quilt Trail. But while those listings let people know the farm was there, they didn't let them see how beautiful it really was. So they made a decision to create a corn maze and open their farm to the public during the fall. This operation quickly grew to include hayrides and a pumpkin patch along with special treats from local kitchens and artists. The rolling hills of the farm make it a beautiful place to relax for a day.

3301 Sable Road, Parrottsville (Cocke County), 423-972-3446, www.neasfarm.com. **f**

Myers Pumpkin Patch and Greenhouse

For three generations, members of the Myers family have raised beef and dairy cattle in Bulls Gap, but they decided to bring the public to the farm when they opened their first greenhouse in 1990. At first, the greenhouse operation was just for flowering plants, but over the past few years, with home gardening's increasing popularity, fruit and vegetable seedlings have become a large part of their spring and summer business.

But fall is really the popular season, when the annual corn maze and pumpkin patch open. Families group around fire pits to cook hotdogs and toast marshmallows, while children play on the farm's purpose-built structures. The variety of pumpkins to choose from is almost overwhelming, but you'll go home with a great one for carving and some delicious winter squash for your table as well.

3415 Gap Creek Road, Bulls Gap (Greene County), 423-235-4796, www.myerspumpkinpatch.com. **f**

Still Hollow Farm

Overlooking the Nolichuckey River, Still Hollow Farm is an idyllic getaway. Groups of students can come out to learn about the history of agriculture. Families can feed the sheep and visit the granary, where all manner of antiques are kept. The grounds are lush with flowers that can be picked, and when fall rolls around, the farm is the perfect place for pumpkins and all your other decorating needs.

3005 West Allen's Bridge Road, Greeneville (Greene County), 423-638-3967, www.stillhollowfarm.com. 🄵

Walnut Ridge Llama Farm

"It's really a llama lifestyle. I guess you could say that we're just llamaholics," Jerry Ayers laughs. That lifestyle is evident when you take a look at the llamas wandering the field and at the store where Carolyn Ayers sells her hand-spun llama yarn and woven products.

For Halloween, the Ayers operate a unique venture—the Spooky Llama Trail. Jerry, a former high school principal, designed the trail with children in mind, and the llamas at Walnut Ridge are accustomed to guests. Instead of being a frightening experience, children learn about Native American legends and folklore. In addition to the Halloween activities, Walnut Ridge offers storytelling and hayrides on Saturday afternoons throughout the fall, and it's a great place to host events year-round.

1345 Chuckey Highway, Chuckey (Greene County), 423-257-2875, www.walnutridgellamas.com.

JEM Organic Farm

Elizabeth Malayter has always been into food. After spending years as a professional chef, becoming a farmer seemed like a good next step. Now, with her husband and daughter, she raises goats, pigs, chickens, and turkeys along with a great variety of produce on their mountain farm.

As a relatively new farmer, Elizabeth admits that she's still learning and that she doesn't ever expect to stop. She has a great philosophy about waste: if something she plants doesn't work, it gets pulled up and fed to her very friendly pigs. She cares a lot about what does work, too. She's heavily involved in protecting heirloom vegetable and fruit varieties and participates in seed-saving projects.

The farm will continue to evolve, but the decision to go organic was never in question. "When we started this, I said, 'I just don't want to use chemicals anymore.'"

385 Sulpher Springs Road, Rogersville (Hawkins County), 423-921-7979, www.jemfarm.blogspot.com.

Ballinger Farm

When you pull into the parking lot at Ballinger Farm's Crazy Maze, one of the first sights you're likely to see is a herd of Holstein dairy cows lazily grazing while they try to figure out what the commotion is all about. And that's exactly what Charles Ballinger wants you to see. "The maze isn't the real draw here. It's the animals and the games and the homemade things we let the kids make. They need to see where these things come from because a lot of them don't know," he says. Ballingers have been farming this land since World War II, raising corn and dairy cows and, until about a decade ago, tobacco. "The corn maze is my son's project. He wanted to do this after we quit growing tobacco. It seems like he wants to go into farming. I just wish he'd chosen something that was less work."

2738 Renfro Road, Jefferson City (Jefferson County), 865-475-7513, www.ballingerfarm.com. ⓕ

Echo Valley Farm

Echo Valley Farm has been in the Simpson family since 1955, with Charlie and Parker Simpson representing the third and fourth generations of the family to farm here, although the farm itself has been in place since 1896. For most of the year, the farm grows row crops of corn and soybeans with fields of hay that feed the dairy cows, but since 2006, the Simpsons have opened the farm to the public in the fall with a huge series of corn mazes and other activities. You can wander the mazes for hours or just enjoy a hot dog roast around a bonfire and take home a pumpkin. If you want to pet farm animals, there are plenty of them here, too, along with a huge play area where even the youngest children can have fun on the farm.

915 Bethel Church Road, Jefferson City (Jefferson County), 865-591-7343, www.echovalleycornmaze.com. ⓕ

Two happy residents of Ward Brothers Farm really dig peanuts.

Ward Brothers Farm

Billy Ward grew up on a farm, and he never gave much thought to organics until his brother was diagnosed with serious food allergies. At that point, the family began using and growing organic products exclusively. Now on his farm, he's looking at sustainability as well as at keeping the farm organic.

The family farm used to be a tobacco farm, and Billy is refurbishing the old tobacco barns to house sheep, chickens, and pigs. When we stopped by for a visit, he was feeding his pigs peanut plant scraps from his fields, where he's also raising an heirloom open-pollinating variety of corn that he'll use as feed for his cattle to supplement the sparse mountain grass when winter comes.

Billy is currently working on his master's degree in sustainable development, so it shouldn't be a surprise that he's always seeing new possibilities for the farm. He's chosen a breed of cattle that is good for beef and milk. His chickens and pigs are good foragers. The sheep he raises now are for meat, but he's talking to sheep farmers and artisans, and wool sheep just may be the next venture for this small but growing farm.

1600 Robe Shull Road, Mountain City (Johnson County), 423-895-3517.

The future of quality Cruze dairy is really cute.

Cruze Dairy Farm

The kids' corn maze at Cruze Dairy Farm is serious business. When they enter, children get to don a farmer's straw hat and carry a bucket. They then "plant" and "grow" corn before getting to "harvest" an ear. Next they shell kernels to feed a "cow," which they have to milk for milk and cream. "Chickens" have to be fed more kernels, and eggs must be gathered. Then the remaining corn is divided into seed for next year's crop and for storage in a silo for the winter. Then things all come together. The kids add their milk, cream, and eggs together in a churn to make ice cream before following the maze to its end and getting a sample of what all their work just created.

Cheri Cruze thought that a corn maze would be a good idea for the farm, and since her husband, Earl, plants a field of corn every year anyway, he decided to listen to her. The corn maze is a great reason to visit this family farm that produces some of East Tennessee's favorite milk. There's also a more complex maze for adults, crafts, and a group of some of the farm's youngest residents, Jersey calves, who are more than happy to have their ears scratched. A visit isn't complete without a stop by the concessions here. Along with daughter Colleen's farm-made ice cream and milk shakes, you can also choose to have some homemade soup beans with fresh corn bread and a pint of Cruze-churned buttermilk.

7309 Kodak Road, Knoxville (Knox County), 865-300-3307, www.cruzefarmgirl.com. **f** 🐦 @CruzeFarm

Oakes Farm

Even on a weekday morning, things are really popping at Oakes Farm. Since 2001, the farm has opened a corn maze to the public, often focusing on local events and heroes as themes. But the corn maze is really just the latest venture for this multigenerational family farm. Oakes Farm is best known for its production of outstanding daylilies, a onetime family hobby that became a successful business. While the corn maze is open only during the fall, you can visit the daylily garden in June and July. Be sure to stop in at the end of June, when the garden is at its peak, for the annual daylily festival.

8240 Corryton Road, Corryton (Knox County), 865-688-6200, www.oakesfarm.com. **f**

Howard Farm

We arrived at Bryant Howard's farm just as the sun went down. He was still showing a steady stream of customers into his eight-acre corn maze, now under a canopy of stars. Some of the bravest were opting to go into the woods for the haunted trail.

When things slowed down for a few minutes, he talked to us about the farm. "It's been here in our family since the 1880s. I'm an animal lover; I'm the director of animal services in Loudon County. But on the farm we raise Angus cattle, Belgians, quarter horses, donkeys. The kids raise sheep, goats, and chickens." The corn maze came about as a way for the family to raise extra money when Bryant's mother became ill, but it's also been a way for the family to meet new people and introduce them to the farm, making sure that they know when they leave that while pumpkins are available at Halloween, the farm has other products that are available year-round.

1675 Malone Road, Loudon (Loudon County), 865-389-6106, www.thehowardfarms.com. **f**

Quae Tae Me Farm

The best part of visiting Quae Tae Me (pronounced "kway tay me") is seeing what goes into maintaining a working farm. You're welcome to participate as much as you want, and Bob and Linda Steppe are happy to share their amazing wealth of knowledge of the land and of the animals that live on it. While we were visiting, we got to meet a fourth-generation farrier and watch him trim the hooves of the farm's three horses.

"Quae tae me" is Gaelic for "come to me," a fitting name for a livestock farm. The entire farm has an Old World feel. Border collies and horses are used to manage the farm's herds of sheep and Irish Dexter cattle that roam the rolling green hills.

1335 Moss Road, Philadelphia (Loudon County), 423-435-3951.

Sweetwater Valley Farm

Cheese! The word always brings a smile to your face for a photo, and the real deal will do the same for you at Sweetwater Valley Farm. Sweetwater controls the entire cheese-making process, from the cows to a retail block of cheese, and visitors to the farm can see the whole thing. If the weather is nice, you can walk down to the barn on a tour to meet the cows and see how technology keeps them on the best possible balanced diet of grazing and feed. Even if you can't take a tour, stop by the Udder Story, the farm's educational exhibit that can teach almost anyone something new about how dairy farms work. And don't miss out on a stop by the retail shop, where you can sample many of Sweetwater's gourmet cheeses and take home some that you won't find anywhere else.

17988 West Lee Highway, Philadelphia (Loudon County), 877-862-4332, www.sweetwatervalley.com. 🅕 🐦 @CheeseCowsWows

Sequatchie Cove Farm

An absolutely beautiful farm nestled among thousands of acres of wilderness is a shining star on the shelves of stores and menus of restaurants up and down Middle and East Tennessee. Members of the Keener family raise pastured cattle, lambs, and hogs for meat and chickens for eggs. They also keep a herd of dairy cattle and produce farmstead cheese. The animals are raised with no hormones or antibiotics, and no pesticides or chemical fertilizers are applied to the land. The Keeners are increasing the fresh vegetables for sale at their farm stand, and when berries are in season, U-pick blueberries and blackberries are available as well.

320 Dixon Cove Road, Sequatchie (Marion County), 423-942-9201, www.sequatchiecovefarm.com. 🅕 🐦 @sequatchiecove

Birchland Ocoee Farms

Joe and Diane Fetzer spend most of the year farming soybeans, corn, and wheat, along with sweet corn, pumpkins, and watermelons, along the banks of the Ocoee River on land that's been in Joe's family since 1904. They open the farm to the public each fall with the River Maze, an experience for the whole family that began as a corn maze and has grown into an all-day agricultural journey. You'll find a soybean maze, compass navigation games, goats to feed, hayrides through the scenic Ocoee valley, and a pumpkin patch, where you can pick one of your own. Going through the corn maze, you'll learn about the importance and history of the Ocoee River.

173 Welcome Valley Road, Benton (Polk County), 423-650-0710, www.therivermaze.com. **f**

Winged Elm Farm

Brian Miller wanted to have a fully diversified farm, but one thing just hasn't worked out. "When we first moved here, I bought a hammock. I hung it up in the trees right over there, and I laid in it once. I've been too busy since," he says. As first-generation farmers, he and his family raise lamb, pork, and beef on seventy acres.

Sustainability is key at Winged Elm. Pastures are grazed in rotation, and each animal is raised as naturally as possible. Pigs are free to root for food in the woods. Steers are raised on grass alone for three years, giving their meat excellent marbling. Rainwater is collected for the animals' use and to fill the ponds on the property. Outbuildings have been built from wood salvaged from the property. It's a system of no waste.

Although Brian still works a full-time job aside from the farm, he sees his family's future in the farm's slow but steady growth.

1285 Sweetwater Road, Philadelphia (Roane County), 865-717-6222, www.wingedelmfarm.com.

Kyker Farms

To say that farming is a long tradition at Kyker Farms is an understatement. "My family has been farming this land for 203 years," Randy Kyker tells us. "It was a land grant from John Sevier—the same John Sevier that Sevier County is named for. We got 174 acres for $174, and we're going on nine generations now." When Randy was growing up, the farm was one of many in the region that raised burley tobacco, but today he raises beef cattle, corn, soybeans, and hay. And he opens his historic farm to the public every fall with a corn maze that he grows and cuts himself every year. "We see people come and then come back again and bring somebody else. I love that."

938 Alder Branch Road, Sevierville (Sevier County), 865-679-4848, www.kykerfarmscornmaze.com. **f**

The Stickley Farm

The Stickley Farm has been in the family since the 1930s, and Debbie Clarke has no intention of letting that change. "There's about 100 arable acres here, and we were leasing that out. But that didn't pay for the taxes, so we had to figure out something else. That's when we decided to start a corn maze to bring people out here." The maze stretches over several acres, and the pumpkin patch, activities, and concessions are located in the scenic hollow beside the antebellum farmhouse. And the corn maze is just the beginning for the Stickley family. "We're planning to build a new barn here for more activities on the farm, and we're going to start a CSA where members can come to the farm to see how we're growing their food," Debbie says.

531 Timbermill Private Drive, Bluff City (Sullivan County), 423-360-4809, www.thestickleyfarm.com. **f** **y** @StickleyFarm

Appalachian Alpacas

Appalachian Alpacas challenges the definition of a family farm. Here, instead of being a farm where family members have grown up and continued a tradition of farming, three sisters have come together to begin that tradition. The Dupont sisters chose alpacas because "we wanted to have animals that we could make money off of without having to slaughter them for meat," Diane Dupont explains. When you spend an afternoon at Appalachian Alpacas, you'll learn all about these fascinating animals; for instance, we learned that, while they're friendly and perfectly willing to eat from your hand, they're also head-shy, meaning that they prefer you to touch their necks instead of their heads.

255 Ralph Rhea Lane, Chuckey (Washington County), 423-257-8110, www.appalachianalpacas.com.

Clover Creek Farm

Chris Wilson started her fifty-acre farm with cattle. Life changed when she got a border collie to help with the cattle and then acquired sheep to help train the dog: she realized that the sheep were a better fit for her small farm. "Sheep are more sustainable," she tells us, "because their little feet don't pack the dirt or tear up the land. And they eat grass and leave manure, so it's a complete cycle."

Chris raises hair sheep, sheep bred for meat rather than for wool. Yes, it sounds backwards, but it means that the sheep have thinner coats of hair instead of heavy wool. Initially, lamb wasn't well received in her community, but now it is very popular, and she has a following of loyal customers who see her every week at the Jonesborough Farmers Market under her banner advertising Green Eggs and Lamb. She's glad to be a part of that because, as she says, "not everyone can raise their own food, but the majority of people can know where their food comes from."

529 Harmony Road, Jonesborough (Washington County), 423-753-2223. **f**

FARM STANDS AND U-PICKS

Blueberry Hill Farm

Blueberry Hill Farm couldn't be more aptly named. After following a winding gravel road to the back of the property, you will find a huge hillside covered in blueberry bushes. Situated just inside the city of Norris, Blueberry Hill will make you feel like you're in a secluded grove a thousand miles away. Owners Brian and B. J. Baxter do expect you to pay for the privilege, but they trust you. You pay for berries here on the honor system, leaving your buckets and payment at the family's barn on your way back to the real world.

101 Reservoir Road, Norris (Anderson County), 865-494-7903,
www.norrisblueberries.com.

Oren Wooden's Apple House

The apple orchard at this third-generation family business was planted in the 1960s as a family orchard. It now covers over 100 acres and is growing every year. The orchard offers over twenty varieties of apples from August through late October. We stopped by on what manager Mark Burnett called "a moderate day." While fresh apples are the orchard's top seller, that day saw a line of people stretched through the Apple House, all waiting for fried pies, doughnuts, cakes, apple dumplings, and more from the pie shop. Since we were there toward the end of apple season, the Apple House was also offering pumpkins and pumpkin goodies, making it a very popular fall day stop.

6351 New Harmony Road, Pikeville (Bledsoe County), 423-447-6376. **f**

R & R Kuntry Pumpkin Center

As the name implies, when fall rolls around, R & R is a great place to find a huge selection of pumpkins, but there's more than that to be found here. Operated by Bobby and Betty Richardson along with their daughter Lisa and her husband, Steve Roberson, this is a true family business. The stand offers produce grown on the family farm all summer, and Betty and Lisa supplement that business with preserves, breads, cakes, and fudge. And the preserves they offer aren't just run-of-the-mill jams and jellies. When we visited, we picked up a jar of beet jelly, something we'd never seen that turned out to be wonderfully tasty. R & R is also the working base for the Tennessee Volunteer Gourd Society, a group that celebrates the history, uses, and culture of gourds. You can see some of the amazing art created with gourds on display at the farm stand.

46901 SR 30, Pikeville (Bledsoe County), 423-447-6352.

Butler's Farm Market

James and Donna Butler's market is especially festive in the fall. Of course, their fresh produce shines, but Halloween was coming up when we dropped in, and the spirit of the holiday was in full swing. You'll find produce here from spring through late fall, along with homemade preserves and local honey and sorghum.

2732 Taylor Road, Maryville (Blount County), 865-984-8435, www.sites.google.com/site/butlersfarmmarket/home. **f**

Coning Family Farm

From late spring through the autumn harvest, produce is abundant at Coning Family Farm, raised by the fourth generation of this farming family. From glorious tomatoes to enormous Halloween pumpkins, you can find it here. While family members often man the farm stand, we felt just as

Honey

When visiting Tennessee farms, you aren't able to meet one of the hardest working groups there—the bees. But you are able to purchase honey, the fruit of their labor, at farmers' markets across the state. What many don't realize is that virtually everything else at the market is there because of bees. While bees are out collecting nectar to make honey, they are also pollinating the plants they visit, including the fruits and vegetables we eat.

Unfortunately, beekeepers are unable to show their hives to visitors. Imagine a territorial bull in a pasture. Now imagine up to 40,000 tiny territorial bulls. Unexpected noises or other disturbances can upset a hive of bees, causing them to swarm. It is possible to get a sneak peek at beehives when beekeepers make occasional visits to fairs and festivals. Beekeeper associations across the state also hold regular meetings that welcome visitors interested in learning more.

much like family with the farm's honor system: the Conings trust their customers to tally up and pay for their purchases.

2724 Taylor Road, Maryville (Blount County), 865-983-0153, www.coningfarm.webs.com. ⓕ

Falls Blueberry Farm

"We love having kids out here so they can learn. Of course they eat a lot while they're out there, but it's worth it," says Bob Falls. Bob will equip you for the perfect two-handed blueberry harvest: a milk jug with the top cut open wide is strapped to your waist by an oversized belt, and you are ready to pick to your heart's content. A lot of practice has gone into creating such a system—Bob's blueberry bushes have been producing for his community since 1983. He readily admits that he could probably make more money from the berries if he had them picked and packaged, but his farm is about the people, especially the ones who come back year after year.

111 Harmon Road, Maryville (Blount County), 865-982-3457.

Maple Lane Farms

The Schmidt family opens Maple Lane to the public twice a year. In spring, U-pick strawberries draw crowds to the farm for the experience of picking berries as much as for the sweet fruit itself. In fall, families come to wander through the Maple Lane corn maze, go on hayrides, and pick pumpkins — and not just a few families. If you visit Maple Lane, you'll be among the over 75,000 visitors that the corn maze draws every season.

1040 Maple Lane, Greenback (Blount County), 865-856-3511.
 @maplelanefarms

Rutherford's Strawberries and Broccoli Farm

Yes. You read that right. Strawberries and broccoli. It's not like you have to eat them together. It's a unique combination for a farm, but it works. Strawberries come in first, with the Rutherford family selling both pre-packed and U-pick. The broccoli starts showing up a few weeks later, but you have to show up early in the morning once it does if you want some. The locals are in the know about just how great Rutherford's broccoli is, so it sells out quickly.

3337 Mint Road, Maryville (Blount County), 865-982-5891.

Apple Valley Orchards

Like many endeavors, Apple Valley Orchards started small. The McSpaddens planted two apple trees in their backyard. Those trees became a hobby that grew into a business with an orchard of over 8,000 trees today. From July through December, the orchard produces apples of over twenty varieties, including Apple Valley's own strain, the Caitlin Gala.

From August through October, during the high season for apples, Apple Valley offers wagon rides and guided tours of the orchard every weekend. And no visit to Apple Valley is complete without a stop by the bakery, where you can purchase fresh apple fritters, turnovers, fried pies, stack cakes, and cinnamon rolls.

351 Reese Road Southeast, Cleveland (Bradley County), 423-472-3044, www.applevalleyorchard.com.

Carver's Orchard and Applehouse

Carver's Applehouse may be off the beaten path, but it's worth seeking out. This farm stand operates on the grounds of a large apple orchard, and apples from the orchard are available year-round. You'll also find other fruits and vegetables in season, along with sorghum and preserves, but the apples are definitely the stars here. Attached to the farm stand is a restaurant where you can enjoy fresh apple cider, apple fritters, apple butter, fried apples, and even apple cider rolls. The menu offers sandwiches and salads as well as classic meat-and-three options.

3460 Cosby Highway, Cosby (Cocke County), 423-487-2710, www.carversappleorchard.com. Dining $

Rowell's Orchard and Motel

The Rowell family owned a motel on U.S. Highway 70 back when parts of the highway were called the Broadway of America. When they learned that the interstate would be coming through, they realized it was time to diversify the business. So they added an apple orchard to the land surrounding the motel.

The Rowells now offer over twenty varieties of apples every year, as well as encyclopedic knowledge of which apples are best for what uses. When asked what Arkansas Black apples are best for, Traci Rowell answered, "It's a hard apple. It's good for cooking and storing. You could use it for self-defense, too. You really could knock somebody out with one of these."

The Rowells have now added peach trees and boast some of the most popular peaches in the area.

6390 Highway 70 East, Crossville (Cumberland County), 931-484-5035.

Ritter Farms

Jack and Nancy Ritter own a farm with a mission. They want people to see where their food comes from, buy it as soon as it's harvested, and enjoy the atmosphere of an old-style farm. Their huge farm stand is in the big red barn, where you'll find heirloom varieties of fruits and vegetables along with baked goods made in-house and canned goods made from the produce grown on their over hundred-acre farm.

The star at Ritter Farms, though, is the Grainger County Tomato. Not only is the tomato a specific variety, but its flavor is enhanced by the unique soil of Grainger County. And the Ritters don't let any tomato go to waste. You'll find plenty of bottled tomato products on their shelves, including spaghetti sauce, chili sauce, salsa, tomato juice, and more.

2999 Highway 11W South, Rutledge (Grainger County), 865-767-2575, www.ritterfarms.com.

Buffalo Trail Orchard

As an engineer, Phillip Ottinger has always enjoyed complex challenges. So it's really no surprise that when he left engineering, he turned to the different, but just as challenging, field of farming. In 2008, he planted a five-acre orchard of eleven varieties of apples, some familiar, others rare, but all scheduled to keep the orchard busy from July through October. October also sees the arrival of Buffalo Trail's annual pumpkin patch with hayrides into the field. In spring and early summer, you can enjoy berries in season: blueberries, blackberries, and both black and red raspberries.

1890 Dodd Branch Road, Greeneville (Greene County), 423-639-2297, www.buffalotrailorchard.com.

Middle Creek Blueberry Farm

Cross the creek and spend an afternoon on the mountaintop filling your buckets with berries from mature bushes on this three-and-a-half-acre farm. You'll be happy knowing that Middle Creek never uses sprays in its berry patches. In addition to blueberries, Middle Creek will keep you picking from April through the first frost with strawberries, blackberries, and raspberries.

595 Middle Creek Road, Afton (Greene County), 423-636-2624, www.stonepile.org/mcbf.

Crabtree Farms of Chattanooga

You're most likely to visit Crabtree Farms as a farm stand, but there's so much more for you to see and do. This amazing urban farm inside the city limits of Chattanooga exists to promote research and education in sustainable agriculture and provides great services like the *TasteBuds* local food guide that promotes local products throughout the region.

The farm offers tours for large groups and also hosts tours for smaller groups, if staff is available. Also, check the farm's website and Facebook

Chattanooga Gaining Ground

The Chattanooga Gaining Ground initiative was launched in 2010 with the goal of increasing the production and consumption of local foods in the Chattanooga area. Gaining Ground provides funding and support to bring people together to form productive and creative partnerships that will further Gaining Ground's goal. Why is this so important? A local food system bolsters the local economy, is better for the environment, is better for the health of the people involved, and builds a sense of community.

Through the generosity of the Benwood Foundation, Gaining Ground has been able to financially advance progress of the Main Street Farmers Market, an important project to bring more local food to an up-and-coming part of Chattanooga. It's also helped Crabtree Farms develop the *TasteBuds* local food guide and website and worked with the Hamilton County Department of Education to bring local food into the school system. In addition, Gaining Ground provides resources for Chattanooga natives and visitors alike. Look to its website for an online local food guide, links to the latest issues of *TasteBuds*, and a downtown local food map.

page before planning your visit; there are often special events and volunteer days at the farm when you can participate in a more hands-on fashion.

1000 East 30th Street, Chattanooga (Hamilton County), 423-493-9155, www.crabtreefarms.org. 🟦 🟦 @Crabtree_Farms

Fairmount Orchard

Since 1928, Fairmount Orchard has been providing a variety of apples to the Chattanooga area. Located on Signal Mountain, the orchard offers a beautiful view as well as delicious products. Open from September through the holiday season, you'll be drawn inside the shop by the aroma of hot cider. Try a glass of freshly pressed cold cider. And of course you can take some home.

2204 Fairmount Pike, Signal Mountain (Hamilton County), 423-886-1226. 🟦

Melody Orchard

This U-pick orchard will keep you busy from spring through fall with berries, peaches, and apples all season. But there's more here than delicious fruit. You'll find a bakery and café offering fresh-baked goods paired with huge servings of ice cream and also a shady patio where you can sit down to enjoy them.

102 Blevins Road, Rogersville (Hawkins County), 423-345-2426, www.melodyorchardtn.com. ▪

Black Oak Farms

While the huge peach orchard is the focus of Black Oak, the farm also offers berries in early summer, sweet corn in late summer, and muscadines in the fall. This is a beautiful farm with a lovely pond that will make you want to stay and pick fruit all day.

7235 Corryton Road, Corryton (Knox County), 865-687-6900. ▪

Clear Springs Farm

At its farm stand, this mountain farm offers seasonal berries, vegetables, and herbs, all naturally grown. When they're available, you can purchase potted herbs, greens, native perennials, berry canes, and blueberry bushes as well as horse manure tea to nourish your own garden.

While buying produce may be your primary reason to visit the farm, it won't be the only reason that you'll come back. Plenty of free-roaming chickens, ducks, and guinea hens wander around the yard and are accustomed to interacting with customers, children in particular. There are hiking trails and a creek to wade in that make this a perfect place to have a picnic. Just remember to bring something extra to share with those guard guineas.

8411 Thompson School Road, Corryton (Knox County), 865-622-0380, www.clearspringsfarm.org.

Fruit and Berry Patch

Dennis Fox, PhD, has a jovial disposition and a beard worthy of Santa Claus. Originally a professor of food science, it was his beard that led him to start his U-pick operation. His boss had been giving him grief over what was then a neatly trimmed professorial beard—and then an opportunity came for a change. "This was my father-in-law's land. One day I was walking here, and the idea struck me. It was clear as day." He now grows a wide variety of fruits and berries along with his beard. The kitchen on-site uses fruit from the orchard to make fried pies, preserves, cider, fruit slushes, and barbecue sauce.

4407 McCloud Road, Knoxville (Knox County), 865-922-3779, www.thefruitandberrypatch.com.

King's Hydrofarm

When you think of farms, you probably think of rolling green hills with rows of crops or herds of animals. King's Hydrofarm takes a different approach. Owner Janet King says, "As more pressure is put on farmland, farmers will have to grow up, not out." The Kings are doing just that. Using hydroponics and hoop houses, they are able to grow seven acres of plants on just a quarter-acre of land. In addition to saving space, hydroponics also means using less water and fewer chemicals. Whatever produce their customers don't pick, Janet turns into preserves, and the proceeds from one special garden are donated to charity.

3238 Tipton Station Road, Knoxville (Knox County), 865-406-7454, www.kingshydrofarm.com.

Mayfield Farm and Nursery

Yes, the Mayfields of Mayfield Farm and Nursery are the same Mayfields who were behind Mayfield Dairy before it was sold in the 1990s. But just because Mayfield Dairy is a large operation doesn't mean that this isn't a family farm. In fact, it's just the opposite. Mayfield Farm sits on the land that the Mayfield family first settled in 1820, and the present-day Mayfields use it primarily to raise fruits and vegetables that are sold in their on-site farm market, through a CSA program, and at area farmers' markets. The farm also has a special area set aside in the fall for a corn maze, pumpkin patch, and other activities. And dairy operations may be gone from the farm, but they're not forgotten. A milking parlor is maintained on-site, mainly as an educational experience for children's groups that come to visit.

257 Highway 307, Athens (McMinn County), 423-746-9859, www.mayfieldfarmandnursery.com. ⓕ

Shultz Farm

This Tennessee Century Farm produces beef and grain for most of the year. But for the past twenty-five years, fall has brought visitors to the farm for apples, pumpkins, and more. The Shultz family will give you a tour of the farm, including a wagon ride out to the fields. In the farm stand, you can pick up different heirloom varieties of apples as they come into season, along with pumpkins and gourds.

For those who want more than just raw apples or pumpkins, the farm stand stocks dried apples, fried apple pies, and fresh apple cider. You can also grab some of the farm's apple butter and honey from a local beekeeper who brings his bees to pollinate the apple trees. And if you're going to be in the area during the holiday season, be sure to give Cecelia Shultz a call for one of her homemade dried apple stack cakes, a traditional Appalachian treat.

245 County Road 603, Athens (McMinn County), 423-745-4723.

Bollenbacher's Blueberries

You'll have to do a little bit of work to locate Bollenbacher's Blueberries, but if you follow the signs, you'll find a shady nook hiding a beautiful patch of mature blueberry bushes just waiting to be picked. Gustav Bollenbacher tells us that he hasn't used insecticides on his bushes in over thirty years, and these are large bushes just bursting with fruit. Stop by the house for your bucket and then follow the twisting drive up the hill to the berry patch, where you're sure to come back with a full bucket and stained lips.

445 Old Sweetwater Road, Sweetwater (Monroe County), 423-337-9562.

Strawberry Knob Farms

Come ready to work when you visit Strawberry Knob. This small farm near Madisonville is a great place for U-pick strawberries every spring, but you'll discover pretty quickly that strawberry picking isn't as easy as it looks. It's worth it for these berries, though. They're small, sweet, and perfect for eating from the bucket, baking into muffins or pies, or turning into preserves.

3250 New Highway 68, Madisonville (Monroe County), 423-836-1133, www.strawberryknobfarms.com.

Tsali Notch is unbeatable for its vineyard views.

Homemade pepper jelly glows in the sun at Delano Community
Farm Market.

Tsali Notch Vineyard

Our two favorite things when we visited Tsali Notch were the muscadine
juice and the beautiful view. The thirty-five acres of muscadines are all
on gently sloping hills. Stand in the middle of the vineyard, and the view
seems to go on forever. On those thirty-five acres, owner J. D. Dalton grows
five varieties of this native southern grape. Some of the muscadines go to
local wineries and home winemakers, while others go into the vineyard's
own bottled juice and jellies. But not to worry, there's still plenty left on
the vine for you to pick.

140 Harrison Road, Madisonville (Monroe County), 423-506-9895,
www.tsalinotch.com.

Delano Community Farm Market

The absence of electric lights and the use of a mechanical cash register
distinguish this market. The Mennonite community members who run the
market avoid technology as much as possible. They do, however, raise excellent produce. Although their beliefs may look to the past, their farming
is forward-thinking as they grow a tremendous variety of fruits and vegetables. In season, this is the best place to find the community's own heirloom
tomato, the Delano Green Tomato, along with other hard-to-find heirloom
vegetables like zipper cream peas and small stringless sweet potatoes. This

is an exceptional opportunity to experience a different world. The workers are generally quiet, but they are willing to answer any questions you may have about the produce, community, or Mennonite life.

146 Needle Eye Lane, Delano (Polk County).

Baxter's Orchard

Since the 1930s, Baxter's has offered over twenty varieties of apples. You can choose as many or as few as you want, mixing and matching the available varieties. Along with fresh apples, you'll find homemade jellies and jams, local honey, country ham, and peanut brittle. The orchard used to press cider from its own apples. But since it doesn't have pasteurization equipment of its own, it now sells cider from a neighboring orchard.

5446 East Parkway, Cosby (Sevier County), 865-217-2281.

Mountain Mist Farms

Hermione and Harry May operate this huge farm of U-pick blueberries, blackberries, and raspberries, with apple and peach trees beginning to produce fruit. Their well-established bushes grow in a beautiful valley that's close to the tourist mecca of Pigeon Forge but feels far away and peaceful. And keep an eye out for wildlife. While we were there, we came across a beautiful mountain tortoise.

710 Caney Creek Road, Pigeon Forge (Sevier County), 865-258-3276, www.tennesseemountainmistfarms.com. ⬛

Larry Thompson Farms

You'll find yourself visiting the Thompson family several times over the course of the year. In spring, the Thompsons offer beautiful strawberries from their on-site farm stand, and then as summer produce comes in, the stand fills with tomatoes, squash, melons, corn, peppers, cucumbers, and almost anything else you might want. And of course, once fall swings around, there are plenty of pumpkins and winter squash for you to choose from. But you won't be coming here just for great produce. The Thompsons are knowledgeable about what they grow and want to make sure that you'll be happy with what you take home from their farm, so they're prepared to answer questions and offer suggestions.

236 Charlie Carson Road, Jonesborough (Washington County), 423-753-4429, www.larrythompsonfarm.com. ⬛

FARMERS' MARKETS

Oak Ridge Farmers Market
During World War II, Oak Ridge was a center for top secret atomic research. Today, the city is still known as the "Secret City," but one thing is for sure: its farmers' market is no secret. Even on the chilly fall day we visited the town, families flocked to the market to browse the abundance of produce. Men talked politics, and the line for samples of fresh bread never dwindled. This market is one of our personal favorites because of its proximity to the iconic Big Ed's Pizza.

Jackson Square at Georgia Avenue, Oak Ridge (Anderson County), 865-310-8617, www.easttnfarmmarkets.org/oak-ridge-farmers-market.asp. ▪

Maryville Farmers Market
This market sets up in the heart of downtown Maryville. The selection of goods available is impressive when you realize that everything sold here is either grown or made on nearby farms. The market is based on a strong community of family farmers who openly offer one another support. Enjoy live music as you walk through the market, and don't hesitate to try some samples of the wares for sale.

Church Avenue, Maryville (Blount County), 865-696-5107, www.maryvillefarmersmarket.org. ▪

Newport Farmers Market
This small farmers' market is a grassroots effort dedicated to supplying locally grown food to the Newport community. Small doesn't mean limited, though. You'll find strawberries, seasonal produce, flowers, blueberries, breads, preserves, eggs, soaps, and more.

You'll notice that the market takes place in the parking lot of the Tanner Cultural Center. The center is the former Tanner School, the local African American school until the racial integration of schools in the 1960s. After refurbishment, it became the Cultural Center and offers extensive support and cultural activities, including the farmers' market, to the community.

115 Mulberry Street, Newport (Cocke County), 423-623-9272, www.newporttnfarmersmarket.com. ▪

Farming and Hunger

From our personal experience, having a farm also means having a garden to feed your family. Today, however, many people living in rural areas experience hunger on a regular basis. One reason for this is the consolidation of farms into single large corporate entities. Now, people in rural areas are not necessarily farmers and do not have access to significant amounts of land for gardening.

According to Feeding America, there are several additional reasons for the sad fact that there is hunger among people living in communities that provide our food. Unemployment and underemployment tend to be very high in these communities, and the work that is available often pays low wages. There is little infrastructure such as daycare and transportation in place to support working families. Finally, the large demand for food assistance stresses what systems are in place, making it harder to keep up.

Food banks across the nation are working to end hunger in both rural and urban areas. One group of allies they have found is farmers. The Garden Writers Association started the "Plant a Row for the Hungry" program in 1995. Since then, over 18 million pounds of produce have been raised just from people planting an extra row or just an extra plant.

Call your local food bank and ask about donations of fresh produce. If you don't garden, this is one more great reason to start. If you already garden, then you no longer have to worry about what to do with all that extra squash and cabbage. Someone out there would love to have them.

Appalachian Farmers Market

This small market is one of the most welcoming in the region. The vendors here are all locals who are raising and making what they sell. You'll find a good variety of produce, fresh eggs, flowers, baked goods, and crafts. The market's permanent pavilion means that even on the chilly mornings early in the market season, vendors and shoppers have a warm place to congregate.

Fair Street at Lakeview Street, Gray (Greene County), 423-477-3211, www.appfarmersmarket.com. �facebook 🐦 @Appfarmermarket

Greeneville Farmers Market

In its permanent pavilion at the Greeneville Fairgrounds, the Greeneville Farmers Market offers great opportunities for farmers and shoppers. This is a producer-only market, meaning that vendors can sell only what they grow or make, and you'll see that the locals of this area grow and make a great variety for you to choose from. You'll find seasonal produce, specialty meats, fresh eggs, baked goods, preserves, flowers and plants, crafts, art, and more as new vendors join the market every year. The market also supports sustainable living through its ongoing teaching series, the Home Grocery Seminars. These classes teach you how to create a pantry, store food most efficiently, and use the items in your pantry to save time, dollars, and stress.

123 Fairgrounds Circle, Greeneville (Greene County), 423-367-9495, www.greenevillefarmersmarket.com. 🇫

Chattanooga Market

The Chattanooga Market has not one but two excellent locations. The Saturday market takes place next to the Tennessee Aquarium in the heart of Chattanooga's exciting downtown. The Sunday market is in the south end of downtown at the First Tennessee Pavilion, a former foundry that has been updated into an open-air area. In addition to providing a place for shopping and socializing, the market has kept $2 million in the local economy. The vendors here sell only what they produce, and they produce a great variety of meats, dairy products, baked goods, eggs, produce, crafts, and more. Market visitors get to enjoy performances by local musicians, demonstrations by local artisans and chefs, seasonal market festivals, and many educational opportunities for both children and adults.

1 Broad Street, 1829 Carter Street, Chattanooga (Hamilton County), 423-402-9960, www.chattanoogamarket.com. 🇫 🐦 @chattamarket

Miniature tomatoes await shoppers in the early morning mist at the Dandridge Farmers Market.

Dandridge Farmers Market

A small-town market can be just that, small, but the Dandridge Farmers Market is an exception in every way. Here, up to twenty-five vendors line a city parking lot offering only those things they raise or make themselves. We saw heirloom varieties of tomatoes, beans, squash, and more at almost every stand. This is a busy market with a very supportive community providing plenty of room for more growth to come.

Gay and Meeting Streets, Dandridge (Jefferson County), 865-397-3977, www.mainstreetdandridge.com/Farmers_Market.html.

Johnson County Farmers Market

Depending on where you're coming from, the Johnson County Farmers Market may be hard to get to. You'll probably have to follow winding mountain roads to reach this market in Mountain City, but it's worth the drive to buy some of the beautiful mountain produce you'll find here. There's not as long of a growing season at this elevation, so this market doesn't extend much into spring or fall. But when we visited, we were able to enjoy traditional Appalachian music while we shopped and found local treats like mushrooms, freshly baked bread, preserves, fresh eggs, and more.

110 Court Street, Mountain City (Johnson County), 423-895-9980, www.johnsoncountyfarmersmarket.org.

Dixie Lee Farmers Market

This is a young, growing market located in the parking lot of a shopping center. You'll find produce, meat, baked goods, and crafts from local vendors, and you can also enjoy a great breakfast or lunch from the shopping center's restaurants that open early just to support the market.

12740 Kingston Pike, Knoxville (Knox County), 865-816-3023, www.dixieleefarmersmarket.com. **f**

Market Square Farmers Market

Make sure you're wearing your walking shoes when you visit the Market Square Farmers Market in downtown Knoxville. This is a huge farmers' market, stretching over two city blocks with a street fair–like atmosphere. Artists and musicians abound for entertainment, and shoppers come out in force. It's no wonder they do; you can find anything you can possibly imagine at this market, and all of it is produced locally.

Market Square, Knoxville (Knox County), 865-405-3135, www.knoxvillemarketsquare.com/farmersmarket. **f**

New Harvest Park Farmers Market

The excellent facilities at this market make everyday feel like a festival. The park has a playground for the kids and a splash pad that keeps little bodies cool and makes bigger bodies long to join in the fun. There's also a picnic pavilion, and the purpose-built Community Building is fully equipped for classes. The vendors provide a full range of produce, meats, and crafts with special surprises like teas, jerkies, and dog treats. And maybe best of all, you can get dessert and fresh milk or ice cream to go enjoy in the shade.

4700 New Harvest Lane, Knoxville (Knox County), 865-215-2340, www.knoxcounty.org/farmersmarket. **f**

University of Tennessee Farmers Market

It's easy to spend a whole afternoon at the UT Farmers Market. Show up early to peruse the beautiful UT Gardens and enjoy a picnic in the shade before stocking up on fresh produce from local growers and students in the UT Organic and Sustainable Crop Production program. You can also purchase local milk and meat as well as baked goods and seasonal creations from students at the UT Culinary Institute. If that's not enough, the market

Tennessee Agricultural Extension Service

The Tennessee Agricultural Extension Service is part of the University of Tennessee's Institute of Agriculture. The extension has an office in every county in the state, all with the mission of protecting the state's natural resources and improving life for all the state's citizens. The extension reaches out not just to farmers but also to the families and communities of Tennessee. Classes and educational material are offered to teach nutrition, food safety, financial responsibility, and proper parenting.

Farmers can get assistance with water quality protection, farm building and layout planning, soil testing, and much more. Livestock farmers can obtain information on nutrition, waste management, and improvements through breeding. Homeowners can learn how to sustainably improve their lawns and even keep bees.

offers various educational and children's activities and maintains a demonstration kitchen garden to give potential gardeners help and inspiration.

2506 Jacob Drive, Knoxville (Knox County), 865-974-8332, www.vegetables.tennessee.edu/UTFM.html.

Market Park Farmers Market of Athens

The long-term plan for this market is off to a good start with an attractive pavilion and friendly farmers selling a wide range of goods. A second pavilion is in the works, as is an amphitheater and exercise walkways. This market won't just be a place to buy local goods; it will also be a place to spend a day with educational opportunities for adults and children, exercise classes for people of all ages and abilities, performance space for local artists, and more.

South Jackson Street at Highway 30, Athens (McMinn County), 423-744-2704, www.cityofathenstn.com/marketpark.

Gatlinburg Farmers Market

You'll find around twenty vendors at the Gatlinburg Farmers Market in the scenic downtown area, and market manager Jenny Burke is optimistic that the market is just beginning to shine. Shoppers are a good mix of local regulars and tourists looking for fresh produce to stock their cabin kitchens. Everything sold here is locally grown with vendors selling only what they produce. To support the huge area crafts community, the market features a weekly craft demonstration. Local chefs also pitch in with cooking demos that fill the air with delicious aromas and tempt visitors to taste what they can create in their own kitchens.

705 East Parkway, Gatlinburg (Sevier County), 865-332-4769, www.gatlinburgfarmersmarket.com. ⓕ

Kingsport Farmers Market

The Kingsport Farmers Market is a hub of activity on Saturday mornings. This busy open-air market hosts roughly thirty vendors. Fresh produce is the most common offering across the market, but you can also find local meats, artisan baked goods, preserves, and a variety of crafts. When we were there early in the season, many of the vendors were offering seedlings for use in customers' own gardens. Some vendors also offer samples, making sure that you'll be going home with their delicious wares.

Center Street at Clinchfield Street, Kingsport (Sullivan County), 423-357-3897, www.kingsporttn.gov/kingsport-farmers-market. ⓕ

State Street Farmers Market

If you want to push the boundaries of Tennessee agritourism, visit the State Street Farmers Market in Bristol. As you drive down State Street, notice the Tennessee flags on buildings on the south side and Virginia flags on the north—you're on the state line. The market sits on the Tennessee side, and like all border city markets, it offers excellent local produce and meats from the surrounding states as well as from Tennessee. But no matter where the vendors are from, they're all proud to be bringing their products to this lively market in beautiful downtown Bristol.

801 State Street, Bristol (Sullivan County), 423-764-4171, market.bristoltn.org. ⓕ 𝕏 @statestreetmkt

There is no question of producer pride at the Johnson City Farmer's Market.

Johnson City Farmer's Market

This open-air market is one of the most popular places in town on Saturday mornings. And it's no wonder, with the large selection of local vendors here and their array of products. You'll find tables loaded with fresh produce, baked goods, preserves, artisan goods, and more. We almost brought home a baby pig when we visited, so be prepared for surprises.

South Roan Street at State of Franklin Road, Johnson City (Washington County), 423-202-1012, www.johnsoncityfarmersmarket.com. **f**

Jonesborough Farmers Market

In Jonesborough, the oldest city in the state and home to the International Storytelling Center, farmers and their customers take time to chat and exchange stories of their own. The market is held in the historic district of Jonesborough beside the Washington County courthouse. In the heart of this district, you are surrounded by the past; some of the buildings here date to the 1790s.

The market offers customers a broad selection of goods produced by local farmers and artisans. Chef demonstrations are held weekly, and children's activities encourage younger shoppers to get involved by offering shopping tokens only they can use. The relaxed atmosphere is set by the sounds of a small creek that forms a scenic boundary for the market.

105 Courthouse Square, Jonesborough (Washington County), 423-753-5160, www.jonesborough.locallygrown.net. **f**

CHOOSE-AND-CUT CHRISTMAS TREES

Bluebird Christmas Tree Farm

Leo Collins explains Bluebird Christmas Tree Farm in terms that are close to his heart. "We've grown trees here, but we've also grown our family. All of our kids and grandkids learned to work here. Except for Mack, but then he's just one week old today, so next year . . ."

Leo planted his first trees in 1982 with the intention of selling them wholesale. But when his first trees were ready to harvest in 1987, his wholesale deal fell through. That was when he decided to open the farm to the public as a choose-and-cut operation. "We put out some signs, and we only sold twelve trees that first year, but we had fun. It was good to be a part of Christmas for people. The next year we sold twenty-five, and it just kept growing after that. Now we sell around three hundred every year, but two of those original twelve families still come."

Leo doesn't claim to specialize, but Bluebird is one of the only farms in the area that offers larger trees from ten to fourteen feet tall. Raising larger trees takes more time and effort. As Leo says, "It's important to make money, but that's not what it's all about. I do this because I enjoy it."

While you're at the farm, be sure to take some time to warm up by the fireplace in the pavilion where you can enjoy a hot dog and pick up some locally produced jams and honey.

985 Brushy Valley Road, Heiskell (Anderson County), 865-457-5682, www.bluebirdtrees.com. **f**

Little Mountain Tree Farm

Little Mountain is a scenic wonder with lush green trees as far as the eye can see. But the real adventure is the drive to the farm. The city of Pikeville is on the floor of the Sequatchie Valley, and the farm is atop the Cumberland Plateau at the end of a road that switchbacks up what seems to be a nearly vertical rise. The trip is worth it, though, because owner Andy Bickford offers both cut and dug trees along with advice about caring for them.

3186 Griffith Road, Pikeville (Bledsoe County), 423-881-3904, www.lmtreefarm.com.

Roark Tree Farm

Steve Roark began his Christmas tree farm as a college fund for his son. "He's graduated now and is a respiratory therapist, so it worked," he says, smiling. The farm has been in Steve's family for over a hundred years, and he raised cattle and grew tobacco and hay on the land before turning to a crop closer to his career as a forester, Christmas trees.

"It's been the best profit for the investment, but it's more work than people think," he tells us. "You can't let it go without care. You have to keep the trees weeded and trimmed for shape, and you have to keep a watch for pests and disease all the time."

While the trees still cover the hills thickly on the Roark farm, there aren't as many today as in other years. "I'm looking at retiring in a few years, so we're slowing down. But we'll never be completely out of the business; we'll just make it smaller, especially to keep it around for family and friends."

396 Ridge Road, Tazewell (Claiborne County), 423-869-9010.

ARCY Acres

The second Saturday in December is the day Santa Claus comes to visit at ARCY Acres. Thanks to owners Art and Cyndi Landrigan, Santa and Mrs. Claus also attend three events in the area to support children's charities. The rest of the time, Art slips out of his red suit and into coveralls to work on the farm.

Art and Cyndi are now keeping the celebration going year-round with their shop at the farm. She teaches scrapbooking, and he sells train supplies and offers classes on model railroads. It all comes to a head at Christmas, though. A little girl has her picture taken with Santa and Mrs. Claus, and she giggles when Santa booms "Ho! Ho! Ho!" "That. That's why we built all this," Cyndi says.

4439 Blaylock Road, Crossville (Cumberland County), 931-788-0455, www.arcyacres.com.

Cotton and Forestry—Not All Farms Are for Food

We all owe farmers a debt of gratitude for putting food on our dinner tables. Agritourism gives us an opportunity to see how they do it and to thank them for it. There is another part of agriculture that we don't get to visit but that plays an important role in our lives—the part that puts a roof over our heads and clothes on our backs, not to mention the books on our bookshelves and, for that matter, the bookshelves themselves.

Cotton is less important in Tennessee than it once was, but it is still a part of the state seal as well as a $275 million crop. Forestry is much larger, accounting for $21 billion of the state's economy, including lumber, furniture, paper, and all other wood products and their processing. Managed forests also have the added advantage of providing homes for wildlife and removing carbon dioxide from the atmosphere.

House Mountain Christmas Tree Farm

When Zach and Norma Henry looked out over their twenty acres of saplings their first year as Christmas tree farmers, they were each holding a knife and planning to trim each tree by hand.

"Norma asked me, 'How are we going to do this?' and I just told her, 'One tree at a time.'"

And so they did. The view at House Mountain Christmas Tree Farm today is one of beautifully maintained mature trees that are as much a part of a family home as they are a business.

We walked through a few of the 5,000 trees on the farm with Zach, talking about the business and his plans for customers who've been coming to the farm annually for twenty years. His grandson wants to take over. Zach would like to see the farm stay in the family, but he wants his grandson to work the trees on a smaller scale instead of making it a full-time job. "If I had it to do all over again, I think I would. I might do it differently, but I'd still do it," he told us.

When we asked about his future in the business, he answered us with a huge smile and a sly wink. "I might be ready to retire by the time this year's saplings are ready. I'm eighty-two this year, and it'll take them seven years to be ready to cut. So I might be ready then."

6300 Childs Road, Corryton (Knox County), 865-687-0324, www.housemountainchristmastreefarm.com.

WINERIES

Morris Vineyard and Tennessee Mountainview Winery

You'll enjoy a scenic drive through the countryside, passing fields of grazing cows, when you head to Morris Vineyard. This family-operated vineyard boasts over fifty-two acres, half of which are producing fruit, including twelve acres devoted exclusively to muscadines and scuppernongs. The Morrises opened Tennessee Mountainview Winery in 1986, producing wines exclusively from either their own fruit or that of two nearby farms. While the vineyard is lovely to look at, it offers more than just a view. Visitors can venture out to pick their own blueberries, blackberries, raspberries, grapes, and, of course, muscadines and scuppernongs.

346 Union Grove Road, Charleston (Bradley County), 423-479-7311, www.morrisvineyard.com.

Blue Slip Winery

Blue Slip Winery is proud to use only grapes that it grows and fruit from local growers in its small-batch, handcrafted wines. But you won't see grapevines when you visit the Blue Slip tasting room, located in the basement level of a building in Knoxville's Old City district. Going to a tasting there feels a lot like slipping down into a 1920s speakeasy.

105B West Jackson Avenue, Knoxville (Knox County), 865-249-7808, www.blueslip.com.

Tennessee Valley Winery

The view from Tennessee Valley Winery is breathtaking. The winery sits at the top of a mountain high over a valley with grapevines covering the slope to its bottom. Mountains rise in the distance to create a beautiful backdrop for the winery's many seasonal events. On select Saturdays throughout the spring, summer, and fall, the winery hosts Music on the Mountain, a free event that features local musicians and allows guests to picnic on the winery grounds. The winery also hosts an annual Oktoberfest with a meal and music provided.

15606 Hotchkiss Valley Road East, Loudon (Loudon County), 865-986-5147, www.tnvalleywine.com. f

Savannah Oaks Winery

When this Delano winery opened in 2002, owner Betty Davis knew that she wanted to make it a small family business. The only wines that don't use the Davises' own grapes are their cabernet and chardonnay, and their fruit wines are a rarity. Unlike many others, they're completely made with fruit instead of being based on grape juice. The winery grounds are calm and shady even on the hottest summer days and offer the perfect spot for the many musical events and seasonal festivals the winery offers.

1817 Delano Road, Delano (Polk County), 423-263-2762, www.savannah-oaks-winery.com. f

Apple Barn Winery, Hillside Winery, and Mountain Valley Vineyards

These three sister wineries each produce unique wines made on-site from Tennessee fruits. At Apple Barn Winery, apples are the star; one of its signature wines is made from apples grown in the small orchard beside the winery. Hillside Winery focuses on Italian-style and sparkling wines. Mountain Valley Winery produces French-style wines.

While each winery is unique, the three share a common design: the inner workings of each are displayed behind glass windows in the tasting rooms. Throughout the day, hourly tours are offered of each winery.

Apple Barn Winery, 220 Apple Valley Road, Sevierville (Sevier County), 865-453-9319, www.applebarnwines.com. f

Hillside Winery, 229 Collier Drive, Sevierville (Sevier County), 865-908-8482, www.hillsidewine.com. f

Mountain Valley Vineyards, 2174 Parkway, Pigeon Forge (Sevier County), 865-453-6334, www.mountainvalleywinery.com. f

Smoky Mountain Winery

In operation since 1981, Smoky Mountain Winery is the second oldest winery in the state. All of the wines you'll taste here are made with Tennessee grapes, the majority of which are grown on vineyards thirty-five miles north of Gatlinburg.

The winery itself sits in the heart of Gatlinburg in a Bavarian-style building with the Smoky Mountains as a backdrop. The Bavarian theme continues inside with dark wood shelves and stained glass accents.

450 Cherry Street, Suite 2, Gatlinburg (Sevier County), 865-436-7551. ⓕ

Countryside Vineyards and Winery

Since 1991, Countryside Winery has been a strong supporter and supplier of Tennessee wines. While some grape varietals are brought in from out of state, the majority of the grapes it uses are those grown in its own or local vineyards. This family-owned winery offers tastings of its ever-expanding wine list, and tours are available on request.

658 Henry Harr Road, Blountville (Sullivan County), 423-323-1660, www.cvwineryandsupply.com.

STORES

The Market

The Market is a small full-service grocery store, but it has local selections that rival larger stores. The butcher counter offers bison, elk, lamb, and duck. The dairy case holds milk, butter, and eggs. We were there during strawberry season and were happy to see a table full of strawberries, most from a nearby farm. The bakery section has a variety of breads from multiple bakers. The craft beer and growler wall offers selections from well-known breweries as well as from smaller, local brewers from just a few miles away. There's even local soap at The Market.

606 High Street, Maryville (Blount County), 865-977-8462, www.themarketinmaryville.com. ⓕ ⓨ @themarkettn

Mountain View Bulk Foods

Mennonite-owned markets are usually great places to find locally made breads, desserts, and preserves. But Mountain View Bulk Foods goes further. Here, you can buy freshly churned butter, fresh peanut butter, local eggs and meat, and a great variety of cheeses, many in huge wheels. On Saturdays, the market adds something even better. In a booth in front of the store, three girls spend their day making fresh doughnuts. The aroma alone is enough to make your mouth water even before you see the hot doughnuts hanging on rods to allow the vanilla glaze to dry.

7730 Erwin Highway, Chuckey (Greene County), 423-257-6357. **f**

Link Forty One

Just one word about Link Forty One—"Baconage." It's exactly what it sounds like—sausage made from half ground pork and half ground bacon with a dose of sorghum and spice to round it out. Trae Moore and Tom Montague work behind a large window at their shop in Chattanooga's Southside neighborhood. They take the same open approach to their ingredients, using local, humanely raised heritage pork as well as sorghum and spice blends from local companies.

217 East Main #105, Chattanooga (Hamilton County), 423-322-5525, www.linkfortyone.com. **f** 🐦 @link_41

EarthFare

EarthFare is a regional grocery store that stocks a full line of products from around the world, but it is also making its mark locally. The company's "100 Mile Commitment" means it tries to source as many local items as possible. It also means the company will not label anything "local" if the item came from more than 100 miles from the store it is in. The local produce selection is best in the summer, but the milk and cheese selections are good year-round.

EarthFare has locations in Chattanooga, Johnson City, and Knoxville. www.earthfare.com. **f** 🐦 @EarthFare

Rogersville Produce

This full-service grocery store in the downtown historic district offers a great selection of local produce, honey, sorghum, and fresh breads. It's also a good resource for canning supplies so that you can make the most of the best produce of the season.

711 West Main Street, Rogersville (Hawkins County), 423-272-4817.

Everything Mushrooms

Everything Mushrooms is primarily an Internet business selling books, mushrooms and mushroom products, and everything you need to grow your own mushrooms. The business also has a showroom, so folks in and around Knoxville can benefit from the staff's expertise. They not only provide expertise but work with local foragers to provide a safe and trusted market for the plentiful wild mushrooms, including morels and chanterelles, that grow in the region.

1004 Sevier Avenue, Knoxville (Knox County), 865-329-7566, www.everythingmushrooms.com. f y @blackmorel

Just Ripe

At Just Ripe, local is the first thing you'll see when you walk through the door. Each bin of produce is labeled with origin, and if that origin is a local farm, the farm's name is listed, too. The labeling follows throughout the store on meats, dairy, and eggs. Just Ripe started as a food cart in Knoxville's Market Square, and it hasn't lost those roots in its permanent location. The café side offers daily and seasonal specials made with a focus on being healthy and local.

513 Union Avenue, Knoxville (Knox County), 865-851-9327, www.justripeknoxville.com. f y @just_ripe

Three Rivers Market

This locally owned co-op grocery store has been serving Knoxville since 1981. With strict quality standards in place, Three Rivers sources local goods whenever possible; for those goods unavailable locally, they must be naturally grown, organically grown, environmentally responsible, and produced by independent businesses.

As a food cooperative, Three Rivers is truly member-owned and member-driven. With over 4,000 members, the co-op has a real opportunity to change the marketplace to meet its customers' needs. It can also

Allan Benton inspects one of his famous country hams.

pool resources to provide the economy necessary to be a truly local market. Membership in the co-op is voluntary, so you don't have to be a member to shop there.

1100 North Central Street, Knoxville (Knox County), 865-525-2069, www.threeriversmarket.coop.

Benton's Country Hams

One Tennessean's name is showing up on restaurant menus across the country—Allan Benton. Benton's bacon and country hams have become favorites of chefs throughout the South and beyond because of his devotion to quality and tradition. The woodstove at Benton's burns hickory and apple wood twenty-four hours a day, seven days a week, to produce smoked pork beyond compare. Of the 14,000 hams that Benton's makes every year, all of them come from heritage breeds of pigs. Benton's also offers custom smoking for area farmers who bring in their own meat.

Allan laughs off his fame and popularity. "We don't keep secrets. I'll tell anybody how we do it; it's not that hard. Really, anybody with a smokehouse in the backyard can do it."

2603 Highway 411, Madisonville (Monroe County), 423-442-5003, www.bentonscountryhams2.com.

The Apple Barn

It might be easy to dismiss The Apple Barn as a tourist-only destination, but there's a lot to recommend stopping in there. The store section carries a variety of locally produced items, including country hams and cornmeal. Then there's the cider side of the operation. You can watch and actually tour part of the apple-pressing and cider-bottling lines during the day and then take home a jug of freshly pressed cider to enjoy.

230 Apple Valley Road, Sevierville (Sevier County), 865-453-9319, www.applebarncidermill.com. ⨍

Ogle's Broom Shop

David Ogle is a third-generation artisan with a family heritage of broom-making that stretches back to the 1920s. But he's also a farmer; he raises the broomcorn that becomes the brooms, just as he harvests the wood for the handles from the mountain forests around Gatlinburg. The smell when you walk into the shop is fresh and green. No two brooms are the same; each has a uniquely twisting shape that comes from the branch that became its handle.

670 Glades Road, Gatlinburg (Sevier County), 865-430-4402, www.oglesbroomshop.com. ⨍

DINING

The Farmers Daughter

Never go alone to The Farmers Daughter, and always go hungry. Dining here is family-style, and when the food starts coming, it never seems to stop. Your table will be filled with dishes of salads, vegetables, and two meats for all to share, and once you tell your server that you just can't eat any more, you'll be asked what you want for dessert—and trust us, you'll want dessert. Of course, your table will have been well stocked with hot yeast rolls, corn bread, and butter along with a jar of locally produced sorghum, so dessert can easily be an ongoing event.

And that sorghum isn't the only local product on your table. The Farmers Daughter uses as much local meat and produce as possible to feed the crowds who come there to dine. Speaking of those crowds, plan on there being a wait for a table at this popular restaurant. But don't worry; it's worth it.

7700 Erwin Highway, Chuckey (Greene County), 423-257-4650, www.thefarmersdaughterrestaurant.com. $

Farm to Table

The farm-to-table movement has been a boon to farmers. As consumers want to see local foods in stores and restaurants as well as on their tables at home, farmers see increased revenue streams, but with those benefits come added risks and requirements, depending on the location of the table the farmer is sourcing for. A restaurant may require a specific amount of a specific product every week, meaning that for a farmer to supply that restaurant, he has to have devoted space to ensuring that the product will be available. The more restaurants that do business with a farmer, the more reliable he has to make his farm, independent of all of the inherent risks of nature and man that farmers deal with every year. But some tables require even more. We all want to see our schoolchildren eating better food, and farmers are a big part of making that happen. But in large urban areas, a school system may be feeding over 100,000 children three times per day. That means a lot of lettuce, carrots, and tomatoes. It also means a lot of dependence on area farmers to provide large quantities of the required vegetables. The farmers who participate have to think differently about the way they do business than the farmers you meet at the farmers' market do. While they have the same goal, putting food on people's tables, the logistics are actually very different.

212 Market

Opened in 1992 by Maggie Moses and her daughters Sally and Susan, 212 Market has been one of the most forward-thinking restaurants in Tennessee. Offering a fresh food menu in a meat-and-potatoes town was only one of the challenges the Moses family faced. Although downtown Chattanooga is now one of the state's biggest tourist attractions, in the 1990s it was a mostly empty sea of warehouses and closed businesses. Through the years, the restaurant has evolved and strengthened its commitment to its mission of sustainability. Efforts are made to reduce water and electricity usage. Solar panels on the roof are a big part of that.

Of course at a restaurant, the food is the key element, and 212 Market does just as well there as it does in its environmental work. The kitchen turns out a wide variety of delicious items, using as many local ingredients as possible. As a registered dietitian, Maggie is also able to consult on special diets, so the restaurant is able to serve vegan and gluten-free meals.

212 Market Street, Chattanooga (Hamilton County), 423-265-1212, www.212market.com. Lunch $, dinner $$ 🛅 🐦 @212Market

Alleia Restaurant
Run by James Beard–nominated chef/owner Daniel Lindley, Alleia features a casual menu centered on house-made pasta and pizzas from a hand-built masonry oven. The ingredients emphasize local sourcing and high quality. The decor is unique, from its long communal table to an almost surreal tableau of candles and a frozen waterfall of wax drippings.

25 East Main Street, Chattanooga (Hamilton County), 423-475-6324, www.alleiarestaurant.com. $$ 🛅

The Blue Plate
If you're looking for a taste of local food in Chattanooga without spending a lot of money, look no further than The Blue Plate. This downtown diner offers local products at every meal in a funky setting with a gorgeous view of the Tennessee River. Breakfasts are especially great, with the opportunity to sample locally produced bacon, sausage, cheese, and bread. Free-range eggs are available for a nominal up-charge on request.

191 Chestnut Street, Unit B, Chattanooga (Hamilton County), 423-648-6767, www.theblueplate.info. $ 🛅

St. John's Restaurant | Meeting Place
The side-by-side brainchildren of chef/owner Daniel Lindley and co-owner Josh Carter, these two restaurants bring Chattanooga a gourmet take on the freshest possible local ingredients. While St. John's Restaurant is more elegant and formal, Meeting Place offers a more casual atmosphere. Our favorite item on the menu looks not to local farmers but to the hillsides, home to the leek-like ramps that are harvested from the wild for a potato and ramp soup, among other dishes.

1278 Market Street, Chattanooga (Hamilton County), 423-266-4400, www.stjohnsrestaurant.com. $$$ | $$ 🛅

Urban Stack Burger Lounge

Urban Stack offers "killer burgers and manly drinks." It is unable to source local beef, but the restaurant does use hormone- and antibiotic-free Angus. Much of the rest of its menu comes from the Chattanooga area, though. Sweetwater Valley cheese and Benton's bacon are from just up the road, and the buns are from Niedlov's Breadworks around the corner. Some of the most southern ingredients of all, the pimento cheese and the bologna, are made in-house. And the bar those "manly drinks" come from serves liquor and beer from many Tennessee distillers and brewers. Specialty drinks include the Bacon Manhattan, based on bourbon infused with Benton's bacon.

12 West 13th Street, Chattanooga (Hamilton County), 423-475-5350, www.urbanstack.com. $ 🛐 🐦 @urbanstack

Harry's Delicatessen

Ben and Amy Willis-Becker are Knoxville natives who honed their culinary skills away from home before returning to Knoxville in 2010 to open Harry's. They are dedicated to using local ingredients and to making most of their products in-house. They smoke and cure the meats they buy from local farmers, pickle local cucumbers and beets, bake fresh breads, and more. This is a Jewish/Italian deli, offering the best of both traditions, and it's quickly gained a loyal following. One bite of a Reuben made with house-smoked brisket, and you'll be a follower too.

131 South Gay Street, Knoxville (Knox County), 865-566-0732, www.harrysdelicatessen.com. Lunch $, dinner $$ 🛐 🐦 @HarrysDeliKnTn

Nama Sushi Bar

Naturally, sushi restaurants fly in the freshest fish, because that's what their menu requires. There are opportunities to go local, however, and Nama does just that. Diners not ready to take on raw fish can try tempura, and vegetarians can have vegetable tempura. Nama sources much of its produce from Knoxville area farms, giving a hand to the local economy as well as to timid diners.

506 South Gay Street, 865-633-8539; 5130 Kingston Pike, 865-588-9811; Knoxville (Knox County), www.namasushibar.com. Lunch $, dinner $$ 🛐 🐦 @NamaSushi

Sapphire

A part of the rebirth of Knoxville's downtown, Sapphire is in an 1898 building that housed a jewelry store for seventy years. The dishes served sparkle with local ingredients. One of the most interesting is the PB&J Four Ways, house-made peanut butter paired with four different ingredients on fresh-baked bread. After dark at Sapphire, the draw is the music, with live jazz, different live acts, or a DJ, depending on the day of the week.

428 South Gay Street, Knoxville (Knox County), 865-637-8181,
www.sapphire-knoxville.com. Lunch $, dinner $$ 🟦 🐦 @sapphireknox

The Tomato Head

The Tomato Head has been a fixture of Knoxville's Market Square for over twenty years. This casual eatery offers made-to-order sandwiches, pizzas, and more, using as many local ingredients as possible to guarantee the freshest experience. The "soysage" used in its vegetarian items is made in-house, starting from blocks of tofu. Look in the drink case to find a familiar East Tennessee local favorite, Cruze Dairy milk.

12 Market Square, Knoxville (Knox County), 865-637-4067;
211 West Broadway, Maryville (Blount County), 865-981-1080,
www.thetomatohead.com. $ 🟦 🐦 @thetomatohead

Tellico Grains Bakery

Local and seasonal find their way to your table in delicious form at Tellico Grains. Primarily a sandwich shop, this bakery makes the bread for those sandwiches fresh every day in its wood-fired oven. It also offers local cheeses, meats, and seasonal produce.

When we were there, one display case was devoted to strawberry products, with buckets of freshly picked strawberries from within just a few miles taking over the bottom shelf. If you can't make it to Tellico Grains, it can send its products to you via its online store.

105 Depot Street, Tellico Plains (Monroe County), 423-253-6911,
www.tellico-grains-bakery.com. $ 🟦

English Mountain Trout Farm and Grill

Charlie Ford takes the U-pick farm to an entirely different place—underwater. With a little help from the staff at English Mountain Trout Farm and Grill, you can find the ingredients for your dinner at the end of a fishing pole. Trout go quickly from the stocked pond to your pole to the kitchen, where Charlie's mother, Phyllis, will fry or bake your trout.

The restaurant has been open since the early 1960s. Charlie has had it only for six years, but he feels the history. "A man came to visit and brought his five-year-old. He said that he had only been to Tennessee once before, when he was six. All his life he has remembered fishing here, and he wanted his son to get to experience that."

291 Blowing Cave Road, Sevierville (Sevier County), 865-429-5553. $$ 🄵

The Old Mill Restaurant

It doesn't get more local than next door. While not all of the corn and buckwheat are local, the early-eighteenth-century Old Mill harnesses the power of the Little Pigeon River to grind the grains for demonstrations and sales to customers. The output of the mill also goes to the restaurant next door, where it becomes grits, pancakes, hush puppies, and more.

3344 Butler Street, Pigeon Forge (Sevier County), 865-429-2455, www.oldmillsquare.com/restaurant.htm. $$ 🄵 🐦 @oldmillsquare

Troutdale Kitchens

The Troutdale family of restaurants started with the Troutdale Dining Room in Bristol. Company CEO Ben Zandi used his years of corporate experience to improve the business while still focusing on the use of local ingredients. The model has been so successful that a collection of Troutdale restaurants now exists from Virginia to Tennessee to Georgia.

Various locations, 423-217-1201, www.troutdalekitchens.com. 🄵 🐦 @thetroutdale

LODGING

The Farm at Spring Creek

You may spend only one night in the cabin on The Farm at Spring Creek, but you just might leave ready to start your own farm. Owners Bryan Wright and Lauren Turner offer classes ranging from a half-day lesson on canning to a multiday session of work on the farm sufficient to train you for your own agricultural awakening. Of course, you also have the option of just sitting with your feet up, admiring the view, or strolling around the farm, taking in the glory of nature.

424 Spring Creek Place, Greeneville (Greene County), 423-361-6441, www.thefarmatspringcreek.com. $$$–$$$$ 🛈

Nolichuckey Bluffs

It's always great to escape everyday life and run away to a cabin in the woods, and you can do exactly that at Nolichuckey Bluffs. But you can do so much more. Sure, you can relax on a porch in a rocking chair and enjoy scenic views of mountains, trees, and the river. But you can also choose to wander in the English garden, pick fruit in the apple orchard or berry patches, or really get into the experience of farm-to-table gardening.

Owners Pat and Brooke Sadler have made this bed and breakfast a unique agritourism experience. If you like, you can work in the organic vegetable garden that supplies the café. By weeding, planting, or harvesting, you can earn cash or a discount on your cabin rental. But beyond that, you'll become a part of a system. When a pizza comes out of the wood-burning earthen oven with herbs or vegetables that you had a part in bringing to it, nothing else will taste quite the same.

295 Kinser Park Lane, Greeneville (Greene County), 423-787-7947, www.tennessee-cabins.com. $–$$ 🛈 🐦 @nolichuckeycabs

Grandview Mountain Cottages

When you follow the driveway to the main house at this bed and breakfast, you're actually driving down the old stagecoach trail that once led through the Cumberland to Nashville. But instead of the bustle of stagecoaches stopping to water their horses, you now have wide, tranquil walking paths, a combination fishing hole and swimming pond, and superb views of the Smoky Mountain foothills.

The property offers two cottages, each uniquely designed and decorated but both like an old European family country house. When you stay here, you need to pack only clothes. Co-owner Ilaeka Villa designed these cottages to provide everything else you could possibly think of, from straw hats to shade you from the sun, to canvas shopping bags that you can fill at the local farmers' market, to umbrellas in case it rains.

While the bed portion of this establishment is undeniably beautiful, we would be remiss to leave out the breakfast portion. Papa Phil Bourne, Ilaeka's father, spends his days in the kitchen whipping up simple yet elegant meals for guests along with special treats like his dark chocolate chunk cookies, pickles, and seasonal jams, jellies, and glazes made as much as possible from the farm's own produce and from that sold by local farmers.

2392 Possum Trot Road, Grandview (Rhea County), 423-365-4412, www.grandviewmountaincottages.com. $$$ **f**

SPECIAL EVENTS AND ATTRACTIONS

The Museum of Appalachia

While there's no guarantee that agritourism will lead to romance, it's happened before at The Museum of Appalachia, so you never know—it just might happen again. The museum was founded in 1969 by John Rice Irwin to house his collection of artifacts representing Appalachian life. The museum is not a building, though. Instead, it is a sixty-three-acre plot, and nearly thirty structures are on the land as parts of the collection themselves.

Historic Appalachia thrives at the museum. The restaurant serves country cooking for lunch. Animals roam the grounds freely, and demos are given of the crafts and skills that were required for life in Appalachia. The Fourth of July and the Tennessee Fall Homecoming in October are especially big events, with music and other activities.

It was at one of these festivals that a young clog dancer named Sherry met a sorghum maker named Mark. Now Mark and Sherry Guenther are happily married and working together at Muddy Pond Sorghum. While you may not find your sweetie at The Museum of Appalachia, you are assured of a swell time.

2819 Andersonville Highway, Clinton (Anderson County), 865-494-8957, www.museumofappalachia.com. Dining $ **f**

The Cleveland Apple Festival

Since 2002, The Cleveland Apple Festival has celebrated the regional apple harvest and is growing larger every year. Currently, the festival surrounds the courthouse downtown and draws close to 10,000 visitors. This is primarily a family festival, so it's no surprise that there should be a great selection of children's activities. But it is surprising that all of the activities in the Kid's Zone, including pony rides, are included in the price of admission to the festival. Entertainment focuses on regional specialties, including bluegrass and country with clog dancers taking the stage on occasion. And, of course, there are apples—fresh ones for you to take home, a dessert contest you can enter, a pie eating contest if you're up for it, and more treats in the concessions.

Downtown Cleveland (Bradley County), 423-503-4114, www.clevelandapplefestival.org. Held in October. [f]

Autumn Acres

Stephanie Woods earned a degree in agriculture, then returned to school for her master's in education. The corn maze that she runs with her husband, Bo, allows her to combine those two loves.

Children who come to the maze can learn about the life cycle of plants and their importance as food. They can also visit the petting zoo to interact with farm animals. In the largest of the three mazes, kids of all ages can find their way through the maze by answering the questions, many about agriculture, that are found at each checkpoint. A correct answer leads you to the next checkpoint. An incorrect answer leads you to a dead end and a walk back to try again.

There's also a wagon ride out to the eight-acre pumpkin patch, where kids can pick their very own pumpkin. The store on-site allows anyone not wishing to go for a ride to select from a variety of pumpkins, squash, and gourds. The maze and pumpkin patch succeed in making education fun, but just plain old fun abounds as well.

1096 Baier Road, Crossville (Cumberland County), 931-210-7655, www.autumnacres.net. [f]

Homesteads Apple Festival

The Cumberland Homesteads were created as one of the New Deal Communities to help families recover from the Great Depression. The 252 homes here were designed and built using wood and stone from the land around them, but even before the houses were built, the homesteaders began subsistence farms. The families could live in barns until their houses were built, and they prepared for their homes by having chicken coops, smokehouses, and other outbuildings already in use by the time the houses were done. While they were waiting for the houses, they also worked together on community projects, including a school and the Homesteads Tower.

The festival was created as a fund-raiser to help pay the operating expenses of the Homesteads Tower and Homesteads House Museums. Held on the shared grounds of the tower and the still-in-use Homesteads Elementary School, an apple festival might seem an unusual choice to celebrate the heritage of the homesteaders, but in truth, apples were integral to their lives here.

Apple orchards were some of the first things planted by the homesteaders to feed their families in the early days. Driving through the historic district here today, where approximately 200 of the original homes remain, you'll still see apple trees in the yards of many. Apples were important because of their long harvest season, different varieties, and longevity in storage without the need for drying or canning.

At the festival, you'll see plenty of fresh apples that you can take home, and be sure to check out the bake sale table, where local volunteers supply baked and fried apple pies along with cups of cold cider. You'll enjoy music, crafts, storytelling, kids' games, and apple butter demonstrations and have the opportunity to tour the tower museum. There's also a quilt show and plenty of smoked barbecue if you're hungry for more than apples.

96 Highway 68, Crossville (Cumberland County), 931-456-9663, www.cumberlandhomesteads.org/applefest.html. Held in September.

Grainger County Tomato Festival

There's nothing that says "summer" quite like a juicy red slice of perfectly ripe tomato. And there's no place where tomatoes are celebrated quite like Grainger County. Just a drive through the county will show you how important tomatoes are to the local economy. Tomato farms line the sides of the roads, and tomatoes are sold from stands in many front yards. At the festival, you can purchase an amazing array of tomatoes from multiple local growers, including one of the festival's sponsors, Tennessee Home-grown Tomatoes. Local food vendors offer treats like fried green tomatoes, soup beans, apple stack cakes, and fried pies. You'll also find locally made crafts and a large show of antique tractors.

7480 Rutledge Pike, Rutledge (Grainger County), 865-828-8316, www.graingercountytomatofestival.com. Held in July.

Johnson's Sweet Sorghum Festival

This small annual festival hosted in the front yard of The Farmers Daughter restaurant in Chuckey is one to mark on your calendar. You'll get to watch a horse-powered cane press produce sweet juice that's boiled down to syrup in a small evaporator pan over a wood fire, a task that experience makes look far too easy. You'll also get to watch apple butter being made in a copper kettle over an open flame and corn being hulled and ground into cornmeal and grits. There's also a good selection of homemade preserves and fall produce that you can purchase from the Johnson family.

7700 Erwin Highway, Chuckey (Greene County), 423-257-4238. Held in September.

Nolichucky Vineyard

While the grapes at Nolichucky Vineyard aren't there for you to pick, they still make for a beautiful setting in which to wander. These grapes go to local wineries, where they become part of the Tennessee wines made at some of the wineries included here. But the vineyard offers visitors so much more than grapes. It's a wonderful setting for summer evening concerts and the perfect place to enjoy a picnic beside the Nolichucky River.

6600 Fish Hatchery Road, Russellville (Hamblen County), 423-312-6755, www.nolichuckyvineyard.com. @nolichuckyvines

Bush Bean Visitor Center

Although Bush's Best is a nationwide company now, it all began in the family's general store in Chestnut Hill in 1904. The company initially canned tomatoes but gradually expanded its line of products. During the Great Depression, cheap foods like hominy and beans kept many people and the company going. Today, the company has processing plants across the South, but the original store still stands, functioning as a visitor center.

James Garland plays a dulcimer made from a dried gourd.

There you can learn the history of the company and the secrets behind the canning process. And don't miss out on a visit to the Bush's Family Café for a good southern-style lunch.

3901 U.S. Highway 411, Chestnut Hill (Jefferson County), 865-509-3077, www.bushbeans.com/en_US/about_us/visitor_center. Dining $ 🅵

Fiddlers and Fiddleheads Festival

For a real taste of Appalachian history and heritage, go to the Fiddlers and Fiddleheads Festival at Farmhouse Gallery and Gardens. Every year, the farm opens to the public with bluegrass bands, Revolutionary War re-enactors, and Appalachian artists and artisans. Many of the participants are faculty at East Tennessee State University in the Appalachian Studies program who bring their knowledge along with their art.

We met James H. Garland here and learned from him about his work preserving the acetate recordings of late-nineteenth- and early-twentieth-century artists. We were also privileged to watch him demonstrate his musical skill on his handmade gourd dulcimer.

121 Covered Bridge Lane, Unicoi (Johnson County), 423-743-8799, rhost1.zfx.com/OZXmp/fhgallery. Held in April.

Shady Valley Cranberry Festival

We grow cranberries in Tennessee? Really? You're serious?

That was our reaction the first time we drove through the town of Shady Valley and saw signs for the annual festival. There really are cranberries in Tennessee. But the Shady Valley bog is the southernmost native cranberry bog in the United States, if that makes you feel any better. Locals have gathered wild cranberries here for generations, but in the 1930s and again in the 1960s, the bogs were drained to plant more profitable green beans and tobacco. And so, what was once over 1,000 acres of bogland became only 150 acres.

But that's where the folks at the Nature Conservancy stepped in. They saw the danger these wetlands and the life they supported were in, and so over thirty years ago they began a system of preservation and restoration that is making a substantial difference. While the trails through the bogs are open year-round, guided tours are offered only during the festival. And while the bogs aren't large enough to produce cranberries on a harvestable scale, you can find locals selling cranberry plants that you can take home, along with giving away a lot of knowledge about the truly fascinating cranberry.

Highway 133, Shady Valley (Johnson County). Held in October. ▉

International Biscuit Festival

Soft red winter wheat is not as large a crop in Tennessee as it once was, but flour made from it is still a key ingredient in a southern classic, biscuits. The International Biscuit Festival celebrates not only the biscuit but everything that goes along with it. You don't need anything to improve a biscuit, but a slice of country ham or a good pour of sorghum can't hurt. And of course, many think that the best biscuits are made with buttermilk. The festival brings local producers into the spotlight and honors them with one heck of a party.

Market Square, Knoxville (Knox County), www.biscuitfest.com.
Held in May. ▉ ▉ @biscuitfest

The transportation department pauses for a drink at Ketner's Mill Fair.

Ketner's Mill Fair

A member of the Ketner family has been operating a mill in the Sequatchie Valley since 1824. The current mill ceased full-time operation in 1992, but this water-powered gristmill still grinds corn during the annual fair, where proceeds are used to preserve the historic site. Multiple varieties of cornmeal and grits ground on-site are available for purchase during the fair. You can enjoy mule-drawn wagon rides along with music, crafts, and food from local purveyors. If you'd like, take your family canoeing on the river above the dam. When we visited the fair, farm animals were on loan from the Marion County 4-H Club, and the Mazelin family from the Muddy Pond community was demonstrating sorghum-making, with generous samples readily available.

Whitwell (Marion County), 423-267-5702, www.ketnersmill.org. Held in October.

National Cornbread Festival

South Pittsburg is a quiet town overlooking the cool waters of the Tennessee River. Local industry is red-hot, however, as Lodge Manufacturing has made cast-iron skillets and other cast-iron products in the town since 1896. Given that the highest purpose of a good skillet is making corn bread, each year the town hosts the National Cornbread Festival, where the highlight is a cooking contest with cornmeal as the main ingredient. Plenty of vendors are on hand with standard fair food, but you can also find traditional southern dishes like bean soup and greens—both served with corn bread, of course. Kids can play on the midway, and the whole family can learn as Lodge offers factory tours and local farmers demonstrate historic methods of making cornmeal.

Downtown South Pittsburg (Marion County), www.nationalcornbread.com. Held in April. 🅵 🆈 @NCF2010

Mayfield Dairy Tour

You'll see lots of cows in the fields along the way to Mayfield Dairy. And once you get there, you'll learn during a tour of the Athens bottling facility how those cows are part of the dairy's family. You'll spend an hour learning about the history of Mayfield Dairy while you get to watch a little piece of plastic turn into a gallon jug and get filled with milk. It's an almost dizzying process that puts the old days of Laverne and Shirley at the bottling plant to shame. You'll also learn that all of that milk comes from farms within a hundred miles of the bottling facility.

Of course, the best part of the tour is the end, since that's when you get your sample of Mayfield's delicious ice cream. But you don't have to stop at just a sample or even wait for a tour. The Mayfield Ice Cream Shop will be happy to sell you a scoop or few of your choice whenever you're ready.

4 Mayfield Lane, Athens (McMinn County), 800-629-3435, www.mayfielddairy.com. 🅵 🆈 @MayfieldDairy

National MooFest Dairy Festival

As home of Mayfield Dairy, McMinn County has a rich heritage of dairy history, so it seems only natural that the county should be home to the National MooFest. While there are plenty of festival foods, games, and carnival rides, you'll also find homemade baked goods, local crafts, and, of course, an abundance of dairy products. There are also recipe contests, milk-chugging contests for the brave, ice cream eating contests, and more fun for all ages. There's also free transportation to and from the Mayfield Dairy for tours during the festival, so don't miss out on this fun, educational experience.

204 East College Street, Athens (McMinn County), www.nationalmoofest.com. Held in May. **f**

National Muscadine Festival

We were lucky enough to attend the first year this annual festival took place. The festival is a two-weekend event. The first weekend sees the coronation of the Muscadine Queen. The second weekend has its first day at Tsali Notch Vineyard with muscadine-picking demonstrations, buggy rides, free samples, and more. The second day features a parade through Sweetwater along with vendors, crafts, carnival rides, music, competitions, and, of course, everything muscadine.

Downtown Sweetwater (Monroe County), 423-337-6979, www.nationalmuscadinefestival.com. Held in September. **f**

Polk County Ramp Tramp Festival

Every spring, one particular plant garners a lot of attention in the mountains of East Tennessee. That plant is the ramp, a pungent vegetable related to garlic and onions that breeds only in elevations over 3,000 feet. In Polk County, ramps are celebrated over the course of three days—one day to find and dig up ramps, one day to clean and prepare the harvest, and a third day to feed a crowd of up to 800 people a dinner of ramps. While the dinner is the main event, spending time with the devotees of ramps on the other days is an incredibly educational experience. You'll meet people who've been a part of this festival for over thirty years, and they gladly share their stories of the past along with their knowledge of plant and herb lore. You will also be supporting the Polk County 4-H Club. Visitors are always welcome at any point in the festival, and those who are willing to join in the work are deliciously rewarded.

Camp McCrory, Reliance (Polk County), 423-338-4503, www.polkagextension.com/ramppage.html. Held in April. **f**

FFA and 4-H

No matter how clichéd the phrase may be, children really are our future. Two organizations have been educating and encouraging them in the field of agriculture for a very long time: FFA and 4-H.

FFA, Future Farmers of America, was founded in 1928. The founders saw that agriculture is more than just planting and harvesting; they saw it as a science, a business, and an art. Today, FFA recognizes the great diversity within the field of agriculture and works with its members to encourage them in all of their endeavors. FFA has over 540,000 members in the United States, ages twelve through twenty-one, both male and female. In Tennessee, FFA has chapters in almost every county, more than one in some.

Also founded in the early part of the twentieth century, 4-H began as an educational program to help youth in rural areas develop skills in agriculture and home economics. Today in Tennessee, 4-H is the youth development program of the University of Tennessee Agricultural Extension Service, and the focus on education blended with community resources creates a unique environment for learning. Historically, 4-H was a rural program, with participants raising calves or growing fields of corn. The modern program has created opportunities for youth in urban areas as well, with areas of study in computers and environmental awareness and international cultural exchange programs.

Tennessee Strawberry Festival

The Tennessee Strawberry Festival first took place in Dayton in 1947 to celebrate the county's history as a major area of strawberry production. While many strawberry farmers have switched to other crops over the years, the festival has been an annual celebration. When we went, the weather was less than cooperative, but that didn't stop the festival. Whether you're looking for local music, crafts, or just a lot of strawberry goodies to eat, the Tennessee Strawberry Festival offers fun all around town for everyone.

107 Main Street, Dayton (Rhea County), 423-775-0361,
www.tnstrawberryfestival.com. Held in May. ⓕ

Tennessee "Polk Salad" Festival and Pageant

So "poke sallet" or "polk salad"? Which is it? Well, both actually. Poke sallet is the green, while "Polk Salad Annie" is a rock-and-roll song written in 1968 by Tony Joe White about the green. As a child, David Webb sold poke sallet, helped by his mother. He founded the festival as a tribute to her. The highlight of the day is the lunch of pinto beans, corn bread, and poke sallet. Other foods are available as well, and entertainment and crafts abound in the beautiful riverside park.

Harriman Riverfront Park, Harriman (Roane County),
www.rockytopgen.com/polksalad/. Held in May.

Exchange Place Living History Farm

Exchange Place Living History Farm is exactly what it sounds like. This is the site of the Gaines-Preston farmstead from the 1850s, a self-supporting farm, and the buildings here have been restored authentically to their foundations. A visit here will give you a glimpse into the hard yet beautiful life a self-supporting farm meant for those who lived on one.

Today, this is both a hands-on museum and a working farm. Heirloom produce is grown here, tended by volunteers and turned into traditional stews over an open fire in the kitchen during festivals. Through work with the Minor Breeds Conservancy, animal breeds that were common on farms in this area in the 1850s live and work here.

Spend a Saturday afternoon in the past at Exchange Place, or try to be there for one of their seasonal festivals. At the Spring Garden Fair, you'll learn more about and be able to buy some of the heirloom vegetable and herb plants that grow here. The Farm Fest takes you into the heart of sum-

Fresh ramps to take home are a big draw at the Flag Pond Ramp Fest.

mer on the farm with old-time games and fresh, cold watermelon to cool you off. The Fall Folk Arts Festival brings traditional Appalachian music, apple butter–making, and traditional crafts.

4812 Orebank Road, Kingsport (Sullivan County), 423-288-6071, www.exchangeplace.info. f

Flag Pond Ramp Fest

This one-day ramp festival is all about the meal, and as soon as you get to the festival, you'll smell that meal cooking. Here, you'll feast on freshly cooked ramps and potatoes, soup beans, corn bread, bacon, and all of the sides you can think of, including fresh ramps. While you're eating, you can enjoy listening to local bluegrass, gospel, and country bands and watching clog dancers. And, "if you have the nerve," as festival organizers put it, you are welcome to participate in the annual ramp eating contest.

4361 Old Asheville Highway, Flag Pond (Unicoi County), www.flagpond.com/festival/ramp/fest.htm. Held in May.

Unicoi County Apple Festival

For over thirty years now, the Unicoi County Apple Festival has taken over downtown Erwin. Hundreds of vendors provide food and crafts, and the local senior center and local churches offer especially tempting baked goods, pastries, and preserves. You'll find many vendors selling apple butter made the old-fashioned way, by stirring ingredients in a cauldron over a fire for hours. You'll also find local orchards offering fresh apples by the bag, peck, or bushel. And if you've never had it, this is a great place to look for a traditional Appalachian treat, dried apple stack cake.

Downtown Erwin (Unicoi County), 423-743-3000, www.unicoicounty.org/applefest.php. Held in October.

Wayne Scott Strawberry Festival

This festival truly celebrates all things strawberry. All food vendors here represent local nonprofit groups, and many of them, especially the senior center, focus on strawberry dishes. You'll also find arts and crafts, classic cars, and family-friendly activities like a pie throwing contest, a cornhole tournament, and a train for the kids. You may also find some surprises. In addition to all of the strawberry options, one vendor was selling authentic Mexican-style street food when we were there.

404 Massachusetts Avenue, Unicoi (Unicoi County), www.unicoicountytn.gov. Held in May.

RECIPES

Gazpacho Cold Summer Soup with Heirloom Bell Peppers and Tomatoes

At Grandview Mountain Cottages, Papa Phil tries to keep meals light in the summer heat. One of his favorite summertime traditions is the classic gazpacho cold summer soup with freshly baked bread. Nutritious and smacking of flavor, Papa Phil uses heirloom veggies straight from his garden. Why heirloom? For a variety of flavor you just can't find otherwise.

SERVES 6

5	heirloom tomatoes, divided
2	cucumbers, divided
1	small sweet red onion, chopped
1	garlic clove, minced
¼	teaspoon sea salt
¼	teaspoon freshly ground black pepper
1	teaspoon fresh basil, chopped (green and purple if available)
1	tablespoon red wine vinegar
1	tablespoon freshly squeezed lemon juice
1	cup low-salt tomato juice
1	teaspoon extra virgin olive oil
2	cups seeded heirloom bell peppers, diced

Roughly chop 2 heirloom tomatoes and 1 cucumber.

In a blender or food processor, combine the chopped tomato and cucumber, onion, garlic, salt, pepper, basil, vinegar, lemon juice, tomato juice, and olive oil, processing until a smooth puree forms.

Cover and chill the puree for 4 hours.

Dice the remaining 3 tomatoes and cucumber.

Stir the diced tomato, cucumber, and bell pepper into the puree and serve in chilled bowls.

Autumn Barley Salad

At Just Ripe in Knoxville, Kristen Faerber has access to beautiful seasonal products from throughout the Appalachia region. While you may not have access to the varieties of apples that make this salad unique every week during the season or to the Sweetwater Valley Farm white cheddar that she uses in the café, you'll still have a wonderful salad that makes a delicious and nutritious side on your autumn table or a tasty lunch all by itself.

SERVES 6

FOR THE BARLEY:

1½	cups pearled barley
1	teaspoon safflower oil
1	teaspoon kosher salt

FOR THE TOPPINGS:

3	medium apples, cored and diced into ½-inch pieces
4	teaspoons safflower oil, divided
	Salt and freshly ground black pepper
3	medium carrots, peeled and diced into ½-inch pieces
4	ounces pecans
5	ounces sharp white cheddar, diced into ¼-inch cubes
3	ounces arugula, stems removed and coarsely chopped

FOR THE DRESSING:

¼	cup safflower oil
⅛	cup + 1 tablespoon apple cider vinegar
2	tablespoons sorghum (we use Muddy Pond Sorghum)
	Salt and freshly ground black pepper to taste

Preheat oven to 425 degrees.

TO PREPARE THE BARLEY:

Rinse the barley in a fine mesh strainer.

Heat the safflower oil in a saucepan over medium heat. Add the rinsed and drained barley to the pot and add the salt. Stir the barley around the pot as it toasts, releasing a nutty aroma, about 2 to 3 minutes.

Add 3½ cups of water and cover the pot with a lid, slightly vented to let steam escape. Bring the water to a gentle simmer. Do not stir the barley as it cooks (this will help the grains to remain independent for the salad). The barley will soften in about 20–30 minutes, and the water will be absorbed. The barley will maintain a chewy texture even when fully cooked. Turn the barley out into a large bowl and let it cool while you assemble the rest of the ingredients.

While the barley cooks and cools, roast the apples and carrots, make the dressing, and prepare the pecans, arugula, and white cheddar.

TO PREPARE THE TOPPINGS:

Spread the diced apples on a large sheet pan. Drizzle about 2 teaspoons of safflower oil over them and sprinkle with salt and pepper. Mix the apples and seasonings together with your hands or a spatula and spread them back out on the pan.

Spread the diced carrots on another sheet pan. Drizzle the remaining safflower oil over them and sprinkle with salt and pepper. Mix the carrots and seasonings together with your hands or a spatula and spread them back out on the pan. Roast the apples and carrots in the oven until just softened, stirring a couple of times during roasting. The carrots will need a few minutes longer than the apples, about 10 minutes for the apples and 12 for the carrots. Allow the carrots and apples to cool on their sheet pans.

Spread the pecans on a sheet pan and toast in the oven for about 3 minutes, while the apples and carrots are roasting. They will release a nutty aroma when they are toasted. Set the pecans aside to cool and then chop them coarsely.

TO PREPARE THE DRESSING:

Combine the safflower oil, apple cider vinegar, sorghum syrup, and salt and pepper to taste to make the dressing. The easiest way to blend a dressing is to combine the ingredients in a jar and close tightly with a lid. Shake the jar vigorously to combine the ingredients. Alternatively, you can whisk the dressing together in a large bowl. Taste and adjust seasoning as necessary.

TO MAKE THE SALAD:

After the apples and carrots have cooled, add them to the large bowl containing the cooked barley. Add the diced cheddar and chopped arugula. Add the toasted, chopped pecans. Stir together, then add the dressing. Stir until everything is well combined. Taste the salad and adjust the vinegar, salt, and pepper to taste.

Mild Goat Curry

From her JEM Organic Farm, chef-turned-farmer Elizabeth Malayter is bringing an interesting ingredient to East Tennessee: goat. While goat meat isn't new to the area, it is gaining in popularity. Elizabeth is able not only to provide a good product but to help people use it. This is one of her favorite recipes for the lean red meat.

SERVES 4

½	cup all-purpose flour
¼	teaspoon salt
¼	teaspoon freshly ground black pepper
1	pound goat stew meat
2	tablespoons curry powder
¼	cup olive oil, divided
1	large yellow onion, chopped
½	teaspoon hot pepper flakes (or ½ seeded and diced jalapeño, fresh)
4	medium bell peppers, assorted colors if possible, seeded and chopped, divided
2	tablespoons minced garlic
1	cup red wine (optional)
2	cups chicken or vegetable broth
2	bay leaves
1	tablespoon fresh oregano leaves
1	teaspoon cornstarch

TO SERVE:
Cooked basmati or other fragrant rice
Diced chives or sliced scallions

SPECIAL EQUIPMENT:
Slow cooker

In a medium mixing bowl, combine the flour, salt, and pepper. In a separate mixing bowl, toss the goat meat with the curry powder, being sure to coat all pieces as evenly as possible. Transfer the meat to the mixing bowl containing the flour mixture and toss to coat lightly.

Heat 2 tablespoons of oil in a large skillet over medium heat. Working in batches if necessary, cook the goat until brown, turning several times, about 8 minutes.

Transfer the goat to a clean plate and reserve. Scrape the skillet well, leaving the scrapings in the pan.

Heat an additional teaspoon of oil in the skillet over medium heat. Add the onion and hot pepper flakes or jalapeño. Cook the onions, stirring frequently, until brown, about 5 minutes. Add wine or broth if the onions begin to stick to the pan.

Add half of the bell peppers and the garlic to the onions. Cook, stirring frequently, until the peppers have softened, about 8 minutes.

Add the reserved goat meat, wine, if using, and broth to the onion mixture, stirring to combine, and bring to a strong simmer, about 15 minutes.

While the goat mixture is heating, set the slow cooker temperature to high.

Carefully transfer the goat mixture to the slow cooker and cover. Keep the temperature set to high for 15 to 20 minutes or until bubbles form around the edges of the pot, stirring every 5 minutes.

Turn the slow cooker setting to low and cook, covered, for 2½ hours, stirring occasionally.

Stir in the bay leaves, oregano, and reserved bell peppers.

Ladle 2 cups of the hot broth from the mixture into a small saucepan. Bring the broth to a boil over medium heat. In a small bowl, whisk together 1 teaspoon of cornstarch with 1 tablespoon of water until the cornstarch is dissolved. Add the cornstarch mixture to the broth and cook until thickened, about 5 minutes.

Stir the thickened broth into the slow cooker mixture until the curry reaches desired thickness. Taste and adjust seasoning if necessary.

Serve the goat curry over rice and garnish with diced chives or sliced scallions.

Lazy Wife Pie

Dennis Fox of the Fruit and Berry Patch in Knoxville provided this recipe, also known as "Three Cup Pie," that his mother taught him. "Mama used to make this all the time because she could have it in the oven in less than ten minutes. I've been having this ever since I was just a little tyke."

SERVES 8

1	stick butter
1	quart diced fruit, canned, fresh, or frozen
1½	cups sugar, divided
1	cup self-rising flour
1	cup milk

Preheat the oven to 350 degrees.

Heat a 12-inch cast-iron skillet over medium heat.

Add the butter and stir until the butter has melted and is frothy.

Add the fruit and ½ cup of sugar, stirring to combine. Reduce heat to low.

In a separate bowl, combine the flour, 1 cup of sugar, and milk, stirring to combine.

Pour the batter over the fruit in the skillet. Do not stir the pie after adding the batter.

Transfer the skillet to the oven and bake for 30 minutes or until the batter is golden brown.

Country Ham and Sorghum Caramel Popcorn

Two of the most iconic flavors of East Tennessee are the smoky saltiness of country ham and the dark sweetness of sorghum syrup. This recipe was inspired by two of the producers we've introduced you to in this book, Muddy Pond Sorghum and Benton's Country Hams. While you don't have to use their products to get great results, the combination they create is something uniquely Tennessean.

4	ounces country ham, minced
1	stick (8 ounces) unsalted butter
1	tablespoon bacon grease (optional)
1½	cups sorghum syrup
6	cups popped popcorn

In a saucepan over medium heat, cook the country ham until slightly crisped, about 5 minutes. Transfer the ham to a paper towel–lined plate to drain.

With the heat still on medium, add the butter and bacon grease, if using, to the saucepan. Allow the butter to melt, stirring frequently, about 3 minutes.

Add the sorghum syrup to the butter and cook, stirring constantly, until the syrup mixture comes to a rolling boil. Continue cooking, stirring constantly, until a caramel forms that is thick enough to coat the back of a spoon or to leave a clear path when a spoon is dragged through it along the bottom of the pan, about 10 minutes.

Remove the caramel from the heat.

Pour 3 cups of the popcorn into a large mixing bowl. Add half of the crisped ham and drizzle half of the caramel over the popcorn. Using tongs or buttered hands, toss the popcorn to blend the ham and caramel through evenly. Add the remaining popcorn, ham, and caramel and repeat.

Allow the caramel popcorn to rest for 5 minutes before enjoying.

AGRICULTURE 101 SOME FARM TERMS DEFINED

Agritourism Agritourism is an alternate revenue stream for farmers. Farmers open their farms for tours or sell their products on-site. Agritourism is also an opportunity for people to connect to nature and to the source of their food.

Aquifer An aquifer is an underground layer of permeable rock or of an aggregate like sand or gravel. Water is drawn from aquifers through wells. Aquifer water is generally of high quality because it is filtered by the material that it passes through.

Biodynamic farming Biodynamics is a holistic approach to agriculture. The idea is that all aspects of the farm are integral parts of the whole. Like organics, biodynamics avoids chemical pesticides and fertilizers. Instead, the emphasis is on natural fertilizers like compost and manure. Beneficial insects are used to eat pests, and certain plants are used to drive them away.

Burley tobacco Burley is a light tobacco that is air-dried and used primarily for cigarettes.

Canning A method of food preservation where the food is processed and sealed in an airtight environment. While canning is more labor-intensive than freezing, properly canned food does not require refrigeration until the jar has been opened. Canning classes are often available through agriculture extension offices and farmers' markets.

Century farm A century farm is land that has been owned and farmed by the same family for 100 years or more.

Certified organic Organic produce and products are grown, raised, or made following standards defined by the U.S. Department of

Agriculture. Producers are certified by one of a number of accrediting organizations.

Compost tea Compost consists of organic matter that has broken down. Leaves, grass clippings, and other plant matter become rich nutrients to add to soil in gardens. Compost tea is made by soaking compost in water. Some of the nutrients in the compost are absorbed by the water, which can then be used for plants.

Co-op Also known as a cooperative, this is where farmers pool their resources in certain areas of activity. Supply co-ops allow groups of farmers to take advantage of volume discounts to bring down the costs of seeds, fuel, and farm machinery. Marketing co-ops are often an effective way for small rural farms to get their products to market by having the products managed by a single representative. Other marketing co-ops focus on a specific product in a region, allowing multiple growers to have their products packaged and sold under one brand name.

Corn maze A corn maze is an agritourism attraction where a maze is cut into a field of corn or sorghum. Mazes are cut when the plants are still small. The desired pattern is shaped using GPS. The shortest path through a maze can generally be found by answering trivia questions posted at each fork in the maze.

Cotton gin A cotton gin is a machine that separates cotton fibers from the seeds and other debris.

CSA Community Supported Agriculture is a business model for farmers. Rather than selling their meat or produce directly, they sell shares in their harvest. This allows farmers to have guaranteed income to start the growing season. Customers receive a mix of whatever is freshly harvested.

Dry-land farming An agricultural technique used for the non-irrigated cultivation of dry lands. The system is uniquely dependent on natural rainfall and the efficient capture and conservation of moisture.

Fainting goats Also known as myotonic goats. The muscles of these goats freeze for roughly ten seconds when they feel panic. Though this is painless, it typically results in the goat collapsing on its side.

Farmstead cheeses These are small-batch cheeses produced by hand via traditional methods on the farm from the farm's own herds of dairy animals.

Farm-to-fork The societal movement concerned with producing food locally and delivering that food to local consumers.

FFA The National FFA Organization promotes and supports agricultural education among youth. FFA is one of the largest youth organizations in the United States, with over 540,000 members.

Forestry The science, art, and craft of creating, managing, using, and conserving forests in a sustainable manner.

4-H The 4-H clubs began as afterschool agricultural programs. The four Hs stand for head, heart, hands, and health. The program was created in part because it was thought children would be more receptive to new agricultural technologies through the club's activities. Today, 4-H activities have expanded beyond agriculture, but the focus is still on leadership and youth development.

Gourds Hard-shelled fruits related to the squash family. Dried gourds are often carved or painted for decoration, but they can also be hollowed for use as musical instruments, utensils, birdhouses, and more.

Grits Coarsely ground corn prepared by cooking one part grits in three parts boiling water, usually served with butter and seasoned with salt or sugar.

Heirloom varieties Fruit and vegetable cultivars that were commonly grown during earlier periods of history but that are not grown commercially today on a large scale. They are commonly grown in home gardens and on small farms.

Heritage livestock breeds Much like heirloom varieties of fruit and vegetables, heritage livestock breeds are those breeds that used to be common on family farms but have grown less common with increased industrialization of farming. Heritage breeds encourage diversity and produce animals that are often particularly well adapted to specific local environmental conditions.

Holler A small valley or dry streambed, sometimes referred to as a hollow. "Holler" is the traditional Appalachian pronunciation.

Humanely raised livestock Animals that are raised with specific care to their quality of life. The Certified Humane Raised and Handled program sets standards for each stage of livestock management for meat, dairy, eggs, and poultry.

Hydroponics A method of growing plants using mineral nutrient solutions in water without soil. Plant roots may be only in the mineral solution or in an inert medium such as gravel.

Muscadines and scuppernongs Grapevine species native to the southeastern United States that have been extensively cultivated since the sixteenth century. Ripe muscadines range from dark purple to black, while scuppernongs have a greenish or bronze color. Both varieties can be eaten fresh and are used in wine making, juices, and jellies.

Naturally grown Products produced without synthetic herbicides, pesticides, fertilizers, antibiotics, hormones, or genetically modified organisms. Naturally grown products are technically organic, but their producers have elected not to go through the process of becoming certified organic by the USDA. The Certified Naturally Grown program offers a different option for small farms with potentially less paperwork and lower cost.

Poke sallet or polk salad A plant, usually considered a weed, that is broadly distributed in edge habitats by berry-feeding birds. While the seeds are highly toxic, the juice of the berries is often cooked into a jelly. The leaves of young plants are eaten as a cooked green after repeated blanching, often served with scrambled eggs. Young shoots can also be blanched and eaten as a substitute for asparagus.

Ramps Wild leeks found in Appalachia. Ramp festivals take place in early spring, and ramps are traditionally served fried with potatoes and bacon or scrambled with eggs. Pickled ramps are also common.

Rotational grazing A system of grazing in which herds are regularly and systematically moved to fresh pasture with the intent to maximize the quality and quantity of forage. Resting grazed land allows the vegetation to strengthen, producing pastures with less need for supplemental feed when they are grazed in later cycles.

Row crops Commercial crops, grains, and vegetables that grow well in long rows designed to maximize the benefits of fertilizers and pesticides. Common row crops in Tennessee include wheat, corn, cotton, and soybeans.

Seed-saving The practice of saving seeds or other reproductive material for use from year to year. Seed-saving is the traditional way that farms and gardens were maintained in the pre–industrial age. Today, seed-saving is one way that enthusiasts not only save but also propagate heirloom varieties, by sharing seeds with fellow gardeners.

Sorghum syrup The cooked down, thickened juice pressed from the canes of sorghum plants. Sorghum syrup is a traditional sweetener in Appalachia and much of Middle Tennessee.

Subsistence farm A farm with the primary purpose of producing enough to feed the farmer and his family. Planting decisions are made with that aim in mind and only secondarily toward market prices.

Sustainability The capacity to endure. Agriculturally, sustainability requires an integrated system of plant and animal production practices that will last over the long term, enhance environmental quality, and make the most efficient use of nonrenewable and renewable resources.

Sweet corn A variety of corn with a high sugar content. Sweet corn is harvested in early summer, when it is immature, to be prepared and eaten as a vegetable, as opposed to field corn that isn't harvested until the kernels are dry and mature and is eaten as a grain and used as animal feed.

Urban farm The practice of cultivating, processing, and distributing food in or around a city. Urban farming can include horticulture, animal husbandry, or aquiculture.

COUNTY-BY-COUNTY LISTINGS

Putnam County

Apple Crest Farm 77
DelMonaco Winery 104
Hurricane Hollow Apple Orchard 77
Mark 4 Christmas Tree Farm 98
MMKM Produce at Cockspur Farm 77
Rocky Point Tree Farm 98

Rhea County

Grandview Mountain Cottages 180
Tennessee Strawberry Festival 191

Roane County

Tennessee "Polk Salad" Festival and Pageant 191
Winged Elm Farm 142

Robertson County

Adams Garden 78
Carr Ranch Wild Horse and Burro Center 61
Gammon Family Dairy 62
Pumpkin Place 78
Robertson County Farmers Market 94
Shade Tree Farm 78
Woodall's Strawberries 79

Rutherford County

Blankenship Farms 79
The Blueberry Patch 80
Lucky Ladd Farms 62
Murfreesboro Saturday Market 94
Rutherford County Farmers Market 94
Walden Farm 62

Sequatchie County

Sequatchie Valley Bed and Breakfast Guest Ranch 115
Wheeler's Orchard 80

Williamson County

Amerigo 35
Arrington Vineyards 105
Blackbird Heritage Farm 67
Boyd Mill Farm 85
The Feed Mill 106
Franklin Farmers Market 95
Gentry's Farm 68
Golden Bell Blueberry Farm 85
Hatcher Family Dairy 68
Joe Natural's Farm Store and Café 112
Morning Glory Orchard 86
Real Food Farms 69
Tap Root Farm 70
Triple L Ranch 70
Whole Foods Market 33

Wilson County

Cherry Hill Farms Bed and Breakfast 115
Circle S Farms 87
Fiddler's Grove Historical Village 124
Lester Farms 87
Pumpkin Hill 87
Rhonda and Chris's Tree Land 98

RESOURCES

There is a seemingly unlimited supply of books, magazines, and websites devoted to food and agriculture. Here are some of the ones we find most useful and interesting.

BOOKS

Michael Ableman, *Fields of Plenty* (Chronicle Books, 2005)

Wendell Berry, *Bringing It to the Table* (Counterpoint, 2009)

John T. Edge, ed., *Foodways*, vol. 7 of *The New Encyclopedia of Southern Culture* (University of North Carolina Press, 2007)

Barry Estabrook, *Tomatoland* (Andrews McMeel Publishing, 2011)

Jonathan Safran Foer, *Eating Animals* (Little, Brown and Company, 2009)

Barbara Kingsolver, *Animal, Vegetable, Miracle* (Harper Collins Publishers, 2007)

Marion Nestle, *Food Politics* (University of California Press, 2007)

Michael Pollan, *Food Rules* (Penguin Press, 2011)

———. *In Defense of Food* (Penguin, 2009)

———. *The Omnivore's Dilemma* (Penguin, 2007)

Joel Salatin, *Everything I Want to Do Is Illegal* (Polyface, 2007)

———. *Holy Cows and Hog Heaven* (Polyface, 2005)

Mark Winne, *Closing the Food Gap* (Beacon Press, 2009)

MAGAZINES

Edible Memphis (www.ediblememphis.com)
Hobby Farms (www.hobbyfarms.com)
Local Table (localtable.net)
Urban Farm (www.urbanfarmonline.com)

WEBSITES

National Guides
Local Harvest, www.localharvest.org
Locally Grown, locallygrown.net

Tennessee Guides
Appalachian Sustainable Agriculture Project, www.asapconnections.org
Grow Chattanooga, growchattanooga.org
Grow Memphis, www.growmemphis.org
Middle Tennessee Tourism Council, www.farmfreshfun.com
Pick Tennessee Products, picktnproducts.org
Southeast Tennessee Tourism Association, www.southeasttennessee.com
Tennessee Agritourism Association, tennesseeagritourism.org
Tennessee Christmas Tree Growers Association, www.tennessee
    christmastrees.org
Tennessee Department of Tourism, www.tnvacation.com/agritourism
Tennessee Wildlife Resources Agency Fish Hatchery System,
    www.tn.gov/twra/fish/hatchery/hatchery_sys.html
U.S. Fish and Wildlife Service Dale Hollow National Fish Hatchery,
    www.fws.gov/dalehollow
U.S. Fish and Wildlife Service Erwin National Fish Hatchery,
    www.fws.gov/erwin

Hunger and Nutrition
Feeding America, feedingamerica.org
Garden Writers Association, www.gardenwriters.org
Share Our Strength, strength.org

Sustainability and Environmental Awareness

Certified Naturally Grown, www.naturallygrown.org

SARE (Sustainable Agriculture Research and Education), www.sare.org

Sierra Club, www.sierraclub.org

Southern Sustainable Agriculture Working Group, www.ssawg.org

Tennessee Organic Growers Association, tnorganics.org

USDA National Organic Program, www.ams.usda.gov/nop

GIVING THANKS

Of course, this book would not exist without the kindness and generosity of everyone included in it. We thank them for being so welcoming to us and to everyone who visits them. The role they play in feeding us, educating us, and entertaining us is invaluable.

The Tennessee Department of Agriculture and Department of Tourist Development both do an excellent job promoting our state. Pam Bartholomew and the Pick Tennessee Products program were especially helpful to us when we started looking for people to visit.

The members of the Tennessee Agritourism Association welcomed us with open arms and helped us understand the many different aspects of their work and their lives. We thank them for that and for the work they do.

The University of North Carolina Press is a wonderful home for authors. Our editor, Elaine Maisner, has been a fantastic guide. Our writing and our books are better because of her. We also appreciate everyone who works behind the scenes there, especially Gina Mahalek, Jennifer Hergenroeder, Paul Betz, Caitlin Bell-Butterfield, Beth Lassiter, and Susan Garrett.

INDEX

ABOUT THE AUTHORS

 Paul and Angela Knipple both grew up in Memphis, in families that knew how to farm and spin a yarn. As a child, Paul spent summers at his grandmother's farm in North Mississippi, loving the garden but hating those big green tomato worms. Angela also remembers gardening with her parents and grandparents. Unlike Paul, she loved the tomato worms, along with other creepy crawlies. Her mother learned early on to check her pockets before letting her in the house.

Though Paul and Angela live in Memphis, they still keep chickens in the backyard. They travel extensively, exploring food cultures across the United States and beyond. And they enjoy the amazing variety of cuisines in their hometown. Pairing their love of food with their love of telling stories, they write—both individually and as a team—about where to eat, whom to buy from or eat with, and why what you eat matters. They see food as a social and cultural issue and are committed to the local food movement. Their work has appeared in the *Commercial Appeal*, the *Memphis Flyer*, *Edible Memphis*, *Taste of the South*, *Urban Farm*, and other publications. Their first book, *The World in a Skillet: A Food Lover's Tour of the New American South*, explores the connection between first-generation immigrants in the South and food culture.

You can learn more about Paul and Angela at www.farmfreshtennessee.com.

Other **Southern Gateways Guides** you might enjoy

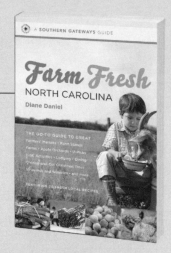

Farm Fresh North Carolina The Go-To Guide to Great Farmers' Markets, Farm Stands, Farms, Apple Orchards, U-Picks, Kids' Activities, Lodging, Dining, Choose-and-Cut Christmas Trees, Vineyards and Wineries, and More

DIANE DANIEL

The one and only guidebook to North Carolina's farms and fresh foods

Wildflowers and Plant Communities of the Southern Appalachian Mountains and Piedmont A Naturalist's Guide to the Carolinas, Virginia, Tennessee, and Georgia

TIMOTHY P. SPIRA

A habitat approach to identifying plants and interpreting nature

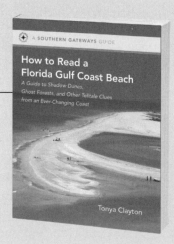

How to Read a Florida Gulf Coast Beach A Guide to Shadow Dunes, Ghost Forests, and Other Telltale Clues from an Ever-Changing Coast

TONYA CLAYTON

A new way to see Florida's dynamic Gulf coast